Barefoot Global Health Diplomacy

Field Experiences in International Relations, Security, and Epidemics

BAREFOOT GLOBAL HEALTH DIPLOMACY

Field Experiences in International Relations, Security, and Epidemics

Author

SEBASTIAN KEVANY, MA, MPH, PhD

Editorial Contributors

DR. AOIFE KIRK, MBBCH BAO

MEGAN HAYES, MSC, BSN, RN

MARCUS MATTHEWS, BA (HONS), MA (HONS), DIP, PGDIP

ELSEVIER

ACADEMIC PRESS

An imprint of Elsevier

Academic Press is an imprint of Elsevier
125 London Wall, London EC2Y 5AS, United Kingdom
525 B Street, Suite 1650, San Diego, CA 92101, United States
50 Hampshire Street, 5th Floor, Cambridge, MA 02139, United States
The Boulevard, Langford Lane, Kidlington, Oxford OX5 1GB, United Kingdom

Notices
Knowledge and best practice in this field are constantly changing. As new research and
experience broaden our understanding, changes in research methods, professional practices,
or medical treatment may become necessary.

Practitioners and researchers must always rely on their own experience and knowledge in
evaluating and using any information, methods, compounds, or experiments described
herein. In using such information or methods they should be mindful of their own safety
and the safety of others, including parties for whom they have a professional responsibility.

To the fullest extent of the law, neither the Publisher nor the authors, contributors, or
editors, assume any liability for any injury and/or damage to persons or property as a matter
of products liability, negligence or otherwise, or from any use or operation of any methods,
products, instructions, or ideas contained in the material herein.

British Library Cataloguing-in-Publication Data
A catalogue record for this book is available from the British Library

Library of Congress Cataloging-in-Publication Data
A catalog record for this book is available from the Library of Congress

ISBN: 978-0-12-818681-7

For Information on all Academic Press publications
visit our website at https://www.elsevier.com/books-and-journals

Publisher: Andre Gerhard Wolff
Acquisitions Editor: Kattie Washington
Editorial Project Manager: Sara Pianavilla
Production Project Manager: Omer Mukthar
Cover Designer: Mark Rogers

Typeset by MPS Limited, Chennai, India

Working together
to grow libraries in
developing countries

www.elsevier.com • www.bookaid.org

Dedication

To the life and work of my father John Kevany who might have laughed at the way in which new cross-over ideas between health, politics, diplomacy, and even war and peace have recently taken off.

To Shannon and Sandy Eng and all of the Eng family for their kind support over many years; for enabling (and helping with recovery from) the missions to, and bizarre experiences in, the far-flung corners of the earth that formed the basis of this book.

To all of those affected by ongoing global pandemics, particularly health-care workers in developing countries: more than others, they know the deep political and international implications of epidemics and infectious disease control.

To everyone in Killiney Bay; to Hugo, Sonny, Rose, Ally, Clem, Peter, Philly, Vicky, Becky, Amanda, Margoe, Raf, T-Boy, Nancy, Pat, Colette, Len, Sophie, Seana, Sabrina, Shane, Tommy, Ferdie, Henry, Stephanie; to the memories of Clemency and James Emmet, Colonel Noel 'Bungo' Craig, and Dr. Yonas Bekele; to the Nimule monkey.

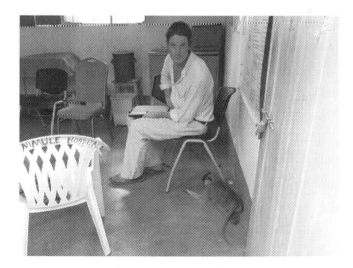

Contents

About the author

Dr. **Sebastian Kevany** has conducted over 100 epidemic response missions to Africa, the Middle East, the South Pacific, Eastern Europe, and Asia with a focus on monitoring and evaluation, health security, international relations, human rights, conflict resolution, diplomacy, and international governance. Sebastian holds BA and MA degrees from Trinity College Dublin (Ireland); an MPH degree from the University of Cape Town (South Africa); is a former adjunct assistant professor at Trinity College Dublin; is an affiliate of the University of California (United States); and holds a doctorate by life research from the University of Westminster (United Kingdom).

The cover picture depicts the author and local medical staff at a remote medical clinic some hours by boat from Taro, Choiseul, and Solomon Islands.

About the contributors

Captain Coral T. Andrews, RN, FACHE (Retired) has served in senior executive roles in the public and private sector. As a military officer, she supported multilateral diplomatic engagements focused on the Asia-Pacific region. In the private sector, she has led the implementation of federal health policy initiatives, served as a state affiliate representative to national trade associations and federal partners, and delivered advisory briefings to the US Government. She holds a BS in Nursing from the University of South Alabama and an MBA from New Hampshire College, and is in her final year of doctoral studies at the University of Southern California (USC). She is also a licensed Registered Nurse and a Fellow of the American College of Healthcare Executives (ACHE).

Dr. Emily Avera is an assistant professor of anthropology at Ithaca College. She holds a a PhD in anthropology from Brown University, MA in anthropology from Leiden University, a M. Phil in diversity studies from University of Cape Town, and BA in politics from Pomona College. For the past decade she has conducted social science research focused on transplant and transfusion medicine. Her current research is focused on blood services in South Africa and has been funded by the Fulbright Program and the National Science Foundation.

Rear Admiral Dr. Michael Baker (Retired) is a general surgeon and trauma specialist. He served 30 years in the U.S. Navy, including combat deployment with the US Marine Corps, and retired with the rank of Rear Admiral. He has authored numerous articles on combat casualty care, strategic medical threats, triage and disaster response, and the use of military in foreign policy which he teaches for University of California Berkeley Osher Life Long Learning Program.

Travis Bias, DO, MPH, DTM&H, FAAFP is a family medicine physician, clinical transformation consultant at 3M, and public health lecturer. He taught medicine for over a year in Kenya and Uganda and has taught courses on global health diplomacy and comparative global health systems for schools of public health in the United States.

Dr. Julia Chang is affiliated with the University of California, San Francisco, and was educated at the David Geffen School of Medicine at the University of California, Los Angeles. Her emergency medicine interests include health services, epidemiology, and global health contexts.

Dr. Sheena M. Eagan is an Assistant Professor with the Department of Bioethics and Interdisciplinary Studies at East Carolina University. Dr. Eagan was awarded her PhD in the medical humanities from the Institute for the Medical Humanities at the University of Texas Medical Branch in Galveston. Her doctoral studies focused on medical ethics and the history of medicine. She also holds a MPH from the Uniformed Services University of the Health Sciences and a BA in philosophy and history from the University of New Brunswick.

Meliza Flores grew up in the Philippines before moving to the United States. She holds a detailed knowledge of the Filipino health system.

Stuart Garrett is an Irish Healthcare Manager and former physiotherapist with an Executive MBA. He has worked and volunteered in several countries including Uganda, India, Nicaragua and Haiti. He is the founder and chairperson of the registered Irish charity "CHEEERS: Developing Healthcare Together" (RCN: 20153335). The charity is a multidisciplinary (MDT) professional volunteer sending organization that recruits healthcare professionals to meet the needs of partner organizations in resource poor settings, focusing on rehabilitation, the growing challenge of non-communicable diseases, and education. Stuart also chairs the Chartered Physiotherapists in International Health and Development (CPIHD), an Irish professional network.

Amy Gidea, RN, has worked for Médecins Sans Frontieres and many other international organizations as a field nurse and in administrative capacities. She is currently affiliated with Coffey International after having lived and worked on HIV/AIDS in Papua new Guinea for some years.

Dianna Kane has spent 15 years practicing design research, strategy and facilitation with global clients and communities. For seven years, she was Chief Design Officer at Medic Mobile, defining strategy through a period of rapid growth and establishing it as a leader in human-centered design. She has also led multi-week, multi-stakeholder design process training sessions in diverse settings for UNICEF, Ministries of Health, foundations and international NGOs. Dianna holds an MPhil in HIV/AIDS & Society from the University of Cape Town and a BA in Urban Studies from Fordham University.

Dr. Alfred Kangole is a research, project and program consultant at the Kampala International University in Tanzania Consultancy Bureau, and was formerly Public Health Specialist and Program Officer for the U.S. Centers for Disease Control and Prevention, Tanzania Office.

Dr. Barry Levine has contributed to e-health projects in Africa, South America, Dominica, and Armenia since 2010. His main focus has been on the OpenMRS medical record system. He has deployed the system, both in the cloud and in local area networks. In addition, he has developed new features and led graduate students on OpenMRS projects. He holds a PhD in Computer Science from Oregon State University. He has served on faculties at University of Maryland, University of Oregon, University of Nairobi, San Francisco State University (SFSU) and American University of Armenia (AUA). He has also served as head of department at SFSU and AUA.

Marcus Matthews is a public policy analyst and independent management consultant with a diploma in investment and finance; a postgraduate diploma in management and corporate governance; and bachelor's degree in politics and philosophy, with a major in geopolitics, terrorism and political communication. Via his master's degree in international conflict and security, he specialized in negotiation and mediation practice, and completed his dissertation on the energy security of the European Union. In the field of global health, he developed an interest in the benefits of global health diplomacy interventions that are supported by the humanitarian application of military structures and their resources, focusing on the political, economic and social return on investment that militaries can provide to both their respective countries and those to which they provide assistance. Marcus believes that the defense policies of countries must evolve to recognize the pivotal role that militaries can play in the combating of complex humanitarian crises, which include not just traditional interventions regarding sudden conflict surges and famines, but also encompass global disasters such as pandemics.

Brian McDomhnaill has worked with the United Nations international Children's Emergency Fund (UNICEF) throughout Africa and Latin America.

Justine McGowan is an independent consultant that has worked for local and international NGOs, UN agencies, universities, and government contractors in the Middle East, Africa, and the United States. She has worked primarily in program management, evaluation, and research in health and democracy programming. Justine holds an AB in Development Studies and Gender and Sexuality Studies from Brown University and an MS in Global Health Sciences from the University of California San Francisco.

Mabvuto Mndau is an international independent health and development consultant based on Malawi. He has worked for Local and International NGOs, UN agencies and private companies in Africa. He has 18 years working experience in programme design, programme management, evaluation, capacity building, institutional systems strengthening, malaria, tuberculosis, health sector reforms, performance based financing, results based management and management. Mabvuto holds a Master of Science degree in Health Services Management (Uganda Martyrs University), Post Graduate Diploma in Programme in Programme Monitoring and Evaluation (Stellenbosch University, Bachelor of Science Degree in Demography (University of Malawi).

Erastus Maina has conducted over 30 global health and development missions in multiple countries in Africa (Kenya, Tanzania, Ethiopia, Sudan, South Sudan) and South East Asia (Cambodia). The focus of these missions was project and investment scoping, risk management, and monitoring and evaluation. Erastus holds BA and MBA degrees from University of Nairobi (Kenya), and an MSc Public Health degree from the University of London (UK). He is a Senior Manager with Dalberg Implement.

Stephen Murphy is Senior Advisor (Donor Relations) at The Global Fund to Fight AIDS, Tuberculosis and Malaria.

Dr. Michele Rubbini is a professor of Colorectal Surgery at the University of Ferrara, Italy. He is a Fellow of American Society of Colon and Rectal Surgeons, and is a specialist in General Surgery. He is also the delegate for the management of the partnership between the University of Ferrara and Namibia University of Science and Technology focusing on the launch of a Bachelor's Degree for Nurses in Namibia. He has published numerous articles on global health and management of research, and is currently working on a new collaborative project between Europe and Africa on e-Health.

Dr. Callum Swift (MBChB. MSc. DTM&H) is affiliated with Tallaght University Hospital, Dublin, and is an emergency medicine doctor with a special interest in global health, wilderness medicine and point of care ultrasound.

Dr Annamarie Bindenagel Šehović is currently Honorary Research Fellow at PAIS, University of Warwick and Associate Research Fellow, Potsdam Center for Policy and Management (PCPM). Her research focuses on health security, in particular on pandemic response and knowledge transfer. Prior to coming to Warwick and Potsdam, where Dr. Šehović was also Acting Professor of International Politics (WS 2017/18), she was a lecturer in international relations at the Willy-Brandt-School of Public Policy / University of Erfurt, Germany. She also spent many years working on HIV and AIDS and health policy in South Africa. She earned her PhD (Dr. rer. pol.) from the Free University Berlin, in 2006. Her most recent publications include: Towards a new definition of health security: a three-part rationale for the twenty-first century, in Global Public Health, and Rethinking Global Health Governance in a Changing World Order for Achieving Sustainable Development: The Role and Potential of the "Rising Powers" Fudan Journal of the Humanities and Social Sciences. She has also written a book on pandemic response, Coordinating Global Health Policy Responses: From HIV/AIDS to Ebola and Beyond (Palgrave, 2017); and another monograph on South Africa's HIV response, HIV/AIDS and the South African State: The Responsibility to Respond. (Ashgate Global Health, 2014).

Karen Weidert is currently the Executive Director of the Bixby Center for Population, Health and Sustainability at the University of California, Berkeley. She has over 10 years of experience in program management, research implementation and evaluation in sub-Saharan Africa. Karen has a diverse background in reproductive health and deep passion for increasing access to family planning worldwide. She received her MPH from the University of California, Berkeley.

Dr Karolina Zielińska, the Institute of Mediterranean and Oriental Cultures, Polish Academy of Sciences, is a graduate of the University of Warsaw specializing in international relations in the Middle East, with particular focus on Israel and its foreign policies. In April 2018 she defended her PhD on Development aid to Sub-Saharan African countries as an element of Israeli soft power.

Acknowledgements

Megan Hayes, Aoife Kirk and Marcus Matthews were all invaluable editorial contributors. The kind and generous contribution and donation of time and effort by all co-authors is greatly appreciated. The guidance of Professor George Rutherford at the University of California, San Francisco throughout the development of these ideas was vital to the production of this book, as was that of Professor Frederick "Skip" Burkle and Admiral Michael Baker. I also acknowledge the contributions of the following: the men of "BOL" for keeping morale high in far out places; Art and Conor and the K-Bay surfers for the missions; Jan and Reed for the San Francisco sessions; to Jer and Sib for their perspectives; to Omer, Sara, and Kattie and all of the Elsevier team in India and the United States who helped to bring this idea in to production. Also Freda Isingoma, Dr. Vinona Bhatia, Dr. Mouplai Das, Professor Ruairi Brugha, Dr. Lauren Kutschzner, Dr. Liz Montgomery, Meliza Flores, Marzena Piertzak, Jay Newberry, Masaw and Francis in Sierra Leone, Payman Jaf, Marisa Dalabetta, Dr. Katie Curran, Dr. Carla Zelaya, Aisli Madden, Tom Osmand, Dr. Benita Panighrahi, Santjie Viljoen, Dr. Carolyn Baer, Christina Spaulding, Dr. Dan Greenwood, Stuart Gaffney, Tammy Chapin, Cristina Delgado, Professor Starley Shade, Professor Carol Dawson-Rose, Dr. Claudia Surjadjaja, Dr. Joshua Michaud, Professor Rebecca Katz, Professor Tom Novotny, Terence Kramer, Matthew Beatty and the "NHL", Christian Beamish, Fatu Yumkella, Dr. Rob Condon, Dr. David Heymann, Dr. David Weakliam, Dr. Jeremy Youde, Jessica Grignon, Mariann Kneisz, Michael Reaume, Professor Niall Roche, Melissa Bingham, Mike Reeves, Dr. Fiona Larkan, Jenniffer Emanuel, Tony O'Mahony, Karin Bridger, Dr. Vishnu Kamineni, Dr. Aidan Hehir, the Banchester Rovers, Barry Dowling, Lindsey Records, Melissa Carnay, Sue Ngo, Amanda Hatfield, Dave and Shane McDonnell, Dave "McLain" Smith, Richie Ghionzoli, Emmet Murphy, Justin Taplin, Ken Ching, Margaret Weir, Dr. Matthew Brown, Dr. Souhial Maleve, Ann Hom, Marianne Outzen, Dr. Brian Vaughn, Dr. Chris Gordon, Willi McFarland, Renee, Sarah Kagan, Dr. Garrett W. Brown, Sheila Baxter, Sophie le Guen, Lucy Kennedy, Kevin Boden, Abbie Laugtug, Vindi Singh, Philly Emmet, Nana Mensah, Marta Romani, Yiuliya Stankova, Maja Barker, Katie Doolan, and Helena King.

Introduction

During the revisions of this manuscript—which were derived from transcripts from a series of university lectures, so my apologies to the reader if the style sometimes seems better suited to the lectern than the page—the world's consciousness of epidemics, pandemics, and infectious disease control underwent some pretty significant changes. From epidemics being vague concepts that affected people far away or long in the past, their virulence and power was brought right to the doorsteps of the most affluent countries in the world.

Global events thus very quickly overtook the composition of this book. Without wanting to make direct reference to it, I quickly realized that a focus on the traditional infectious diseases—HIV/AIDS, tuberculosis, and malaria—would be insufficient. Nor would a focus on key issues such as climate change be suitable, despite its many implications for health.

Everyone has, in some way, now become an amateur epidemiologist: is there a link between weather patterns and the spread of epidemics; are there predisease symptoms? Have government responses been fast enough; have enough geographical details been disclosed of those areas infected—has there been a conflict between the legal, and the medical? Should, say, professions or demographics or vital statistics of those infected have been revealed—and what is the link between the spread of epidemics and other policy events, such as economic restriction?

In all such cases, there has been a link between different professional disciplines and fields of endeavor and global health. This is also happening on the micro level, as individuals never before engaged with the epidemic or infectious disease control realms now come up with innovative ways to engage with the situation themselves—and, one hopes, contribute to its resolution.

Perhaps most importantly, everyone now understands that there is a right and wrong way to respond to epidemics—the wrong way includes pseudoscience; or the advancement of other policies ahead of public health, as many governments around the world have demonstrated. Infectious diseases do not listen to rhetoric, nor to economic forecasts; likewise, excessive liberalism does not seem to be a comfortable complement to pandemics, either. The case has been made for greater authoritarianism, greater central control—up to, and including, the use of the military as a public health instrument. But is that, in realty, any better?

Arguments also of ethics, in economics, in epidemics: economies have rightly been stopped to protect those most at risk, even if they are not in the productive labor force. But have human rights been curtailed in the interest of protecting health systems? Economically, as well—how have the world's billionaires been contributing at this critical time? Is consumer materialism also being challenged, as people for the first time experience limitations on essential supplies—not far flung from rationing in the Second World War? And what about social distancing and compliance with rules as a parallel to blackout policies in World War II? Do these events make the case for population control efforts—for isolationism? And should we really still be calling climate change an "emergency"? The only emergency, it seems, has been the state of the world's health system—and associated inequalities.

The questions go on—how does one balance personal versus social responsibility? Can the rise of neo-conservatism somehow be connected with these events, or is that spurious propaganda? Has tech failed us— why do we have hundreds of different ring tones, but no big data effort that would be programmed to collate the intense groupthink and creative responses coming from around the world? Can epidemics be used to advance selfish, partisan agendas—has this situation been turned into a political tool, or are there many who want a "new normal" anyway? But, isn't it right to use epidemics to advance other lobbying interests, such as banning the live animal trade, just as HIV/AIDS made efforts to reduce social stigma?

What is there also now to say, as well, about the interplay between health security and, for example, neo-isolationism? On the one hand, per-haps there is an appeal—limited mass migration via curtailments on the combined proliferation of low-cost travel and rise of borderless states, leading to relief of housing and environmental issues—and, of course, much greater international health security. On the other hand, disruptions in supply chains, international cooperation, laws of comparative and abso-lute advantage ... sharing of skills, international collaborative expertise ....that can all be threatened, as well.

And the questions still go on .... Are "shelter-in-place" or the more dramatically phrased "lockdown" efforts improving public health in other realms—people staying out of bars and pubs; running and exercising more? Will the environmental gains have been worth it; are we seeing a global return to spiritual consciousness? Either way, amidst all this com-plexity, surely it is telling that so much of the response goes back to John Snow and his pump handle—though it is interesting that the same basic principles apply, it seems staggering that we have to resort to age-old responses in the age of high-tech.

Is this, also, time to examine much more closely world and local leadership both within and beyond health—do we need to be much more certain that we have the most educated and appropriate and highly trained people in positions of political or administrative primacy; people who can speak the language of health, as well as of security and power? Has crucial information been repressed under the banner of protected health data or confidentiality; have governments been truly honest with their citizens in this situation? If not, was that a lesser evil for a greater good?

As I said—so many questions, and the answers will come in time. Yet, as I also said, this isn't a book about that situation. But, it is a book about the interplay between health, epidemics, infectious disease control, public health emergencies, and other things—diplomacy, security, conflict, stability, environment. What hope, for example, is there for the future of humanity if epidemic surveillance and reporting efforts fail on the basis of malign international relations, communications, and diplomacy?

One effect that I thus thought was significant was the intricate web of links that were thus revealed between health and all other aspects of society—economics, politics, the legal, capitalism, authoritarianism, climate change, the environment; culture, social order, international relations, isolationism: those concepts, and many more, were all inextricably linked to health, and often found to have critical connections to epidemics.

In many ways, that is what this book is about: illustrating the relevance of what is called global health to all walks of life, for sure, but also drawing links between other realms—stability, security, conflict, to name but a few—to epidemic threats, and our responses to them. The basic premise is that health programs are worth investing in because not only do they preserve and even sometimes improve the quotient of health in distant countries, but also our own. Not only, in the same way, do they operate as altruistic humanitarianism, but also function as key mechanisms of diplomacy, international relations, and cooperation: this, however, depends in the right program, done in the right way, at the right time.

If such multifariousness of interdigitation brings nontraditional skills to bear on the field of global health—often with unexpected, unmeasured, benign outcomes—then all that remains is to demonstrate these effects. After that, we are left with a powerful motivator for the maintenance or even augmentation of global health, infectious disease control, and epidemic funding—so often classed as expendable—and a line of thinking that is as much of interest to the political rights as to the left, in the enlightened self-interest context.

This book is therefore essentially a political one—yet it also focused very much on the individual. The life blood of the chapters that follow is not so much what I say, which is mainly context and ideas, but the tales from the field provided by a range of correspondents from highly diverse backgrounds—but with one common goal: epidemic and infectious diseases control, and thus the advancement of global health.

These essays demonstrate, in many cases, the key skill sets that medical and other personnel have to develop when engaging in epidemic control efforts in new countries and cultures. This include adaptability, sensitivity, and transparency, as well as many others: consciousness of sustainability, of protocol and etiquette, and of the many benign and malign downstream effects of epidemic control efforts—from their effect on the environment, to whether or not they defy cultural traditions that have survived hundreds of generations.

Of course, health is the key: it is the primary objective of all these other considerations; the beginning, middle and end. But, health doesn't have to be the enemy of the sensitive, the careful, the diplomatic: instead, the right programs (not always the most cost-effective ones) can advance both primary and secondary agendas in unison—synergies, leading to enlightened self-interest. It is to be hoped that the development of these ideas will capture the imagination, not just of the individual—who finds themselves a barefoot diplomat, even though lacking proverbial immunity status—but of countries and institutions, as they begin to understand both the strategic and altruistic effects of, for example, responding to public health emergencies.

In the chapters that follow, we explore these individual level skills, followed by consideration of the need for adaptable programs; locally owned efforts; monitoring and evaluation systems that capture the ephemeral; military and strategic considerations and dividends; and resolving conflicts between ethics and economics that have long plagued epidemic responses.

This is thus a book that sets out how to combine diplomacy, international relations, and epidemic control. It is for field workers—but also for politicians, policymakers, and academics from a range of different disciplines; I hope it will inspire more and more people to understand that their skills and experiences, from almost any background, can be used to advance the world's health—and thereby (when done right) conflict resolution, economic growth, and international relations—and that, in turn, global health is of relevance to all.

CHAPTER ONE

Bare feet: how to do it yourself

Abstract

The theory of global health diplomacy is often challenging to translate into practice, though the preceding chapters have been designed to instill ideas and provide examples of ways in which medicine and international relations can interact in both routine and unexpected ways. For the individual global health worker, either from a clinical or non-clinical background, there are also a set of mindfulness principles that can be extracted from these themes. Through them, everyone involved with epidemic control can gain the rewards and satisfaction of optimizing their contributions not just to the health of the world but also to the stability and security of the global community.

1.1 Diplomatic deficits

"The long labour of peace is an undertaking for every nation. In this effort none of us can remain unaligned; to this goal, none can be uncommitted." [1]

Too often, there are ideas—often dangerously good ideas, such as those voiced by John F. Kennedy, above—that lead nowhere. Ideas that are too esoteric and niche; dead ends and cul-de-sacs: ideas that leave a residual trail of good energy before fading, which are forgotten as quickly as a turned page, or a closed book. So—how about the real? The term barefoot, for example, is evocative of sun and sand, warmth and informality, but also includes *de facto* principles and practices: those who travel around the world in any medical or paramedical capacity, with responsibilities beyond job descriptions, are health diplomats. Barefoot diplomacy thus involves systems and styles to optimize international relations impacts—wherever, whenever.

Barefoot is of course metaphorical: there is no actual need to take your shoes off—unless it is culturally acceptable. Yet barefoot diplomacy requires a level of informality, of humanity— often away from the intimidations of hierarchy; away from shirts, ties and power dressing. Barefoot styles are, instead, a sign of the human, the accessible, and the imperfect: a sign that perhaps beyond profession and purview, you can operate off the grid

Barefoot Global Health Diplomacy.
DOI: https://doi.org/10.1016/B978-0-12-818681-7.00005-4

1

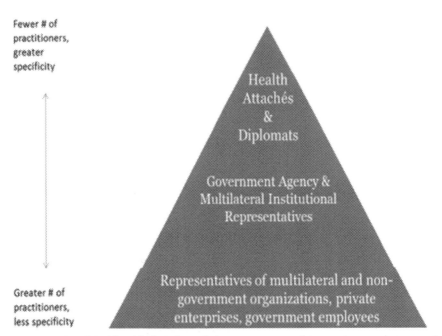

Fewer # of
practitioners,
greater
specificity

Greater # of
practitioners,
less specificity

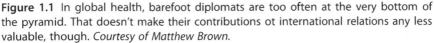

Figure 1.1 In global health, barefoot diplomats are too often at the very bottom of the pyramid. That doesn't make their contributions ot international relations any less valuable, though. *Courtesy of Matthew Brown.*

(Fig. 1.1). A sign also of being in touch with the realities of operating environments; with the visceral and tactile as much as the theoretical and abstract.

Ultimately, and ironically, global health diplomacy creates a formal system for barefoot diplomats. New systems, for instilling new ideas, skills, styles in the infectious disease control world: ways to define the tricks of the trade, a way to merge everyday tasks such as an epidemic control program with conflict resolution. So far, no qualifications or letters after your name are required, only a distinctive style—a sixth sense. Subtle skills, not just for high-profile ambassadors or diplomats but for the low-fi as well: for anyone who wants diplomacy to permeate every aspect of 21st-century life.

Global health professionals are usually already diplomatically inclined—generally they are open-minded, with decent intentions. Yet, they do not always operate and function on broader, non-health levels: too often, they are not ready for it. I was one of those not trained for the role: Instead, I was in the deep end: a month into the job I was on a plane to Zimbabwe and beyond, learning the diplomatic as I went. I was lucky to be travelling with Trudy, a native, who tuned me in to the local politics, the cultural sensitivities, the way to be: taught me about Mugabe, and hyperinflation.

Figure 1.2 In at the deep in end—visiting infectious disease centres in rural Africa. *Picture courtesy Sebastian Kevany.*

I learned the rest as I went, as fast as I could; learning fast, but feeling, in those early days, only one step away from a fatal *faux pas* (Fig. 1.2).

And Zimbabwe was just the tip of the iceberg; it was in fact the start of ten years working on the edge where disease meets war, invasion and natural-resource disputes. Ten years of experiencing the overlap of infectious disease control with realms of fundamentalism and extremism, terrorism and refugees—with extreme religion and dictators, *coups d'états* and Arab Springs; with jihadists, pariah states, police states, revolutions—secessions, and civil wars. It became impossible, and it would I think have been a mistake, to try to ignore or to operate independently of the wider events—the bigger pictures. They were an inevitable part of my work both through influencing programs, and being influenced by them.

A later example was in Uganda, where I saw the government introduce draconian social-cultural legislation. Ironically, that legislation was linked to HIV/AIDS epidemics; connected with global health's new power scope. There, I saw medical workers step up to the diplomatic table, negotiating non-health policy moderation: explaining that stopping the spread of epidemics requires social tolerance, change, and progressiveness. Was that inappropriate involvement in issues beyond health, or was it smart and cool—was it *realpolitik*?

I needed quickly to become a barefoot health diplomat. A capable health diplomat, I read, "must have a sophisticated understanding of structures, programs, approaches, and pit falls to achieve success either in the clinical setting, or at the policy making table" [2]. I also needed to have an understanding of circumstances, the environment, and of bigger pictures: a broader understanding of the principles and practice of international relations and diplomacy, global affairs and priorities, strategies and intrigues: a deeper awareness, which extended far beyond epidemiology. All of this was far from the security training courses that were meant to prepare internationals for anything that comes their way.

Gradually, I learned to understand the role—not only as a *de jure* health care provider, but also as a *de facto* diplomat. Diplomacy had an increasingly explicit purpose; it seemed to be part of personal and professional responsibility. Slowly, I came to understand that in the 20th century, all internationals, implicitly and inherently, held such dual responsibilities: they undertook dual roles, were but usually only recognized for one: if you couldn't measure it, it didn't exist [3]. But, maybe, 21st-century internationals were more and more are being recognized for broader achievements. I felt that recognition, in turn, might enhance global health's profile and credibility, prestige and influence. In Liberia, our first essayist, Dr. Callum Swift, felt the collision between health and international relations both literally and figuratively.

Global health misadventures

Absolute pandemonium. Amongst the chaos, two paramedics run over to us with an unconscious woman, her shirt torn and blood trickling from her ear. Soon, more casualties start arriving. Our radios are crackling with requests for assistance by the front gates, so I put a c-spine collar on the unconscious lady, secure her airway as best I can, and load her into an ambulance, before picking up my grab bag and running towards the entrance. The torrent of people running in the opposite direction make this impossible, and when a paramedic sees me trying he stops me and warns me: *"You cannot go that way doctor, it's too dangerous. Many people are hurt, there are bodies on the ground everywhere, and we can't get to them."* I stand there, bewildered by the shouting and the crying and the heat. It's still early morning, but in Liberia the humidity is oppressive, and sweat and sun cream are stinging my eyes and blurring my vision. This is a crazy place to be.

It's the 22nd January 2018, a historic day for Liberia; the first democratic transition of power since 1944 is about to occur. After a fiercely fought election campaign, George Weah is about to be inaugurated as the 25th president of the republic of Liberia. It's a story for the ages; from a slum dwelling child talent spotted by a local football team, to *Ballon d'Or* winning football hero

(Continued)

(Continued)

and celebrity, and now soon to be president of a nation. Loved by the people, they had been queuing in the thousands since before dawn outside the Samuel K. Doe stadium, and there was a carnival atmosphere in the air.

My journey to that football stadium was a meandering and serendipitous one; whilst still a medical student I had been drawn to Liberia by the promise of uncrowded surf and adventure, so I reached out to the main government hospital, JFK medical centre, and was kindly invited to stay for a one-month elective. After the overwhelming intensity of the hospital, a reprieve was needed, and I headed for the small fishing village of Robertsport. There my brother and I camped on the beach under an ancient cotton tree that has stood sentinel over a series of world class surfing point breaks, since the time the first freed slaves arrived from America to settle. Living on a diet of fresh fish, coconut and empty waves, I lost my heart to Liberia, and knew I would return.

Fast forward a few years, and we found ourselves camping under the same cotton tree, surfing the same perfect waves. The only thing that had changed in the village was the height of the young local surfers that we had met years before. This time, we were just passing through, en route from London to Cape town in a 4×4. I had a job offer back in London starting a few months later, but I had got so used to the rhythm of Africa that the thought of moving to London jarred. Thus I called into the hospital and spoke to the chief medical officer, who said that they would be happy to have me again. I turned down the London job, and after we made it to Cape Town I flew back to Liberia to start work in the hospital's emergency department. (Liberia, if it wasn't already clear, is one of the poorest nations on earth, and the healthcare system is dire.)

I sweated hopelessly in the windowless emergency department against the tide of human suffering, watching young people die from entirely preventable illnesses on a daily basis. But there were enough victories, enough people turned around in the nick of time, to make the work the most rewarding and stimulating I have ever done. And I was privileged to be there at a special time; the Liberian College of Physicians and Surgeons had started a residency training program, and the first cohort of Liberian doctors were graduating from their training, complete with Membership of the West African College of Physicians.

At the graduation ceremony, there was a palpable sense of pride and optimism, for the younger doctors in the audience they had their first role models; not foreign specialists who would invariably leave them after a short while, but Liberian-born and Liberian trained men and women who were now in a position to lead. And for the people graduating, like Emmanuel Ekyinabah, who had given decades of his life as a medical officer, and lived through war and Ebola, the applause and the speeches were vindication of their sacrifice and recognition of their status as Liberia's first homegrown specialists.

(Continued)

(Continued)

Their success was the result of an ongoing residency training program, initaited to try and address the crippling shortage of specialist doctors in Liberia. I'm relatively new to global health and development, but I could see from the start that this program had many excellent characteristics, especially when compared to the plethora of other NGO's and programs I encountered out there. For a start, it wasn't an NGO: expertise and funding came from a variety of sources, from American universities such as Yale to the World Bank—but all of this was channelled through the Liberian College of Physicians and Surgeons. The consultants brought in to train the Liberian doctors were almost exclusively from other African countries, and they were usually on long one- or two-year contracts.

This meant they had time to build up a rapport with the Liberians, who recognized their commitment to their country and acknowledged the similarities between their healthcare systems. These may seem like small things, but they stood in stark contrast to how some of the visiting healthcare professionals acted: flying in for a week, and announcing a string of lectures on topics not discussed with the faculty prior to their arrival (that had often been covered several times already by other visitors), and then giving a talk of which much bore absolutely no relevance to the reality in Liberia. Not surprisingly, the residents' eyes rolled when another of the latter type walked through the door.

Now, finally, the program is playing the long game, investing time and energy into something that took years to start bearing fruit, but that will now pay dividends for generations to come. The foreign consultants will at some point depart, and in their place will be the Liberian men and women who will continue the process of education and growth.

Anyway, the hospital was bustling with rumours of the upcoming change of administration, for Liberia is a small place, and cronyism is rife. The CEO of the hospital was an old friend of the president's, and she knew as well as everyone else that when a new president took over office, her position would be untenable. Decades of her corruption and mismanagement had left the hospital in a worse state when I returned in 2017 than when I had been there in 2013, a remarkable feat given the amount of money pumped into the hospital each year by the government and donors. Not only did the hospital lack many of the medications on the WHO essential medicines list, but also essentials like soap for the operating theatres; my colleagues in obstetrics would often have to walk to the market next to the hospital grounds to buy powdered soap to scrub with. Yet rummage through the many storerooms and outhouses and you could find boxes of unused medications and equipment, including a dialysis machine! A well-meaning but entirely misplaced donation I

(Continued)

(Continued)

am sure—for, with so much work to do to get even the most basic of things done properly, complex treatments like dialysis are a distant dream in Liberia.

One day, I noticed the start of construction work in an area on the ground floor of the hospital—a rarely used thoroughfare that housed neither the hospital entrance, nor any relevant department. On my way to the crumbling and overcrowded emergency department every day, I watched in amazement as this neglected area was renovated into a state-of-the-art lobby, complete with wood panelled ceiling, polished floors, and beautiful carved wooden statues. Nobody I asked seemed to know why so much money was being spent on this particular part of the hospital, and not in the clinical areas. The question was soon answered, however, when it was announced that there would be a special ceremony to mark the completion of the hospital's renovation project—attended by none other than the president herself, Ellen Johnson Sirleaf.

The ceremony took place outside, where the onshore breeze cooled the hundreds of delegates and dignitaries, who sat applauding the rousing speeches about the fine work of the outgoing CEO, Mrs. McDonald. The President thanked her personally, before being shown into the gleaming new lobby. Naturally, there wasn't time for the president to be shown the crumbling wards—but judging by the lobby I have no doubt she was mightily impressed by what she saw, as her acolytes commended the fine investment in the healthcare of the Liberian people.

The disrepair of the hospital mirrored that of the nation, and it was in this context that the upcoming inauguration of George Weah approached. There was excitement and hope in the streets and in the hospital, and I was delighted when I found out that the Minister for Health had requested that JFK Hospital provide a team of doctors for the event, and the chief medical officer asked me to be a part of the team. The keys to a brand-new and fully stocked ambulance that had been parked unused since I arrived were found—another donation of years past. The day before the inauguration we met at the stadium, where the head of the medical operation briefed us about our roles, positions and the locations of the medical tents. The plan was for the twenty or so paramedics to work in zones, and bring any patients to one of two medical tents. There we would assess and treat them, and (if necessary) arrange for transfer by ambulance to hospital for further treatment. The hospitals were to have on-call teams on standby from each of the main specialties, ready to receive patients. The stadium was predicted to be packed, with tens of thousands of jubilant supporters: the temperature was going to be fiercely hot, and nerves were running high. I spent the night anxiously skim reading my copy of the Oxford handbook of prehospital care, poring over the major incident management and mass gatherings chapters (Fig. 1.3).

(Continued)

(Continued)

Figure 1.3 Inauguration Day. *Courtesy of Callum Swift.*

I awoke before dawn, and together with the rest of the team we made our way in the ambulance to the stadium, with what should have been plenty of time. But the traffic was at a standstill, as thousands of people thronged the streets and the area around the stadium; even with the sirens on it took us hours, and we were let through the gates and onto the pitch only ten minutes before the area was due to open to the public.

We had only just started unpacking our gear and setting up the medical tent when pandemonium broke out. As I learned later, a huge crown had been pressing up against the main entrance gate, whipped up with excitement and eager to get front row seats for the spectacle. When the gates were swung open, the crowd charged forward, and as people ran in the maelstrom some fell and were trampled over by the crowd advancing behind them. The screaming started, and the gates couldn't be shut against the tide of people rushing in, so the people lay bloodied and battered on the floor as hundreds of people surged around them—some trying to help them, many running in fear. The crowd then dissipated enough to allow the paramedics to start retrieving the victims and loading them straight into ambulances or bringing them to us in the medical tent, where we worked in a frenzy to triage, assess, treat and transfer them to hospital. As abruptly as the chaos started it ended: the ambulance crews and paramedics did a fantastic job, and the last of around twenty casualties was taken to hospital; the crowds calmed and filled the stadium; and the festival atmosphere resumed.

(Continued)

(Continued)

The rest of the day was uneventful medically, which left us to enjoy our pitch-side seats to the whole affair. The outgoing president and several African heads of state entered to muted applause, dwarfed shortly afterwards by the roar of the stadium as some of Weah's famous football friends from his days in AC Milan and Chelsea entered. Then finally the main act arrived, dressed in a simple white suit and smiling broadly, and the stadium erupted in a jubilant and sustained melee of cheers, drum beats and traditional African song.

This was not his first time playing to an adoring crowd, but as he said afterwards *'I have spent many years of my life in stadiums, but today is a feeling like no other'*. In typical populist style he railed against corruption, unemployment and poverty, but his long and faltering speech was somewhat anticlimactic—and, besides, no one could hear it, so the day rather fizzled out and after the speech the crowds started dispersing.

Today, the Liberian dollar is at a ten-year low, and my friends working in various ministries are depressed and disillusioned. But thankfully, the situation in JFK is a different story altogether, under the new leadership of Dr. Jerry Brown the hospital has seen remarkable improvements. These started literally days after he took over, when he organized a hospital and campus wide clean-up operation, led by himself—complete with wellington boots, rubber gloves and all. The mountains of junk that were piled up on the lawns outside were testament to the decades of neglect the place had been subjected to, and whole operation became both a physical and a spiritual cleansing of the hospital, a chance to start afresh. Since then he has completely renovated the emergency department and several of the wards, as well as finding excellent and motivated replacements for many of the top leadership positions, and there is a palpable sense of change and progress in the air.

I left after a year, to resume my training in Ireland, but there isn't a day that goes by that I don't think of that remarkable country. Recently I helped an Irish colleague go there, and his weekly updates of frustration, joy, amusement, shame, pride and amazement remind me just how intense the experience of being there is. It's an experience that I am sure I will have again, as together with some of our Liberian colleagues we have set up the Liberian Society for Emergency Medicine, and after less than a year of being back in Ireland I am already itching to get back. For there, between the pain and the joy and the exhaustion, I feel that I am sucking on the marrow of life, living in the truest sense—and that is a rare thing (Fig. 1.4).

(Continued)

(Continued)

Figure 1.4 Inauguration Day medical team. *Courtesy of Callum Swift.*

Dr. Callum Swift

1.2 Day-to-day diplomacy

Global health is evolving as a result of these ideas; from a fringe profession—traditionally a profession which was for the missionaries, mercenaries and misfits to one that not only advances the health of the world's poorest people, but also justifies itself to the world by advancing world stability and human rights. Justified also through advancing cooperation, conflict resolution and security, and by becoming a diplomatic art, as well as a health science. This in turn, supports and augments its primary goal: controlling, preventing and treating epidemic diseases.

So, decoding the boundaries of the intersection and overlap between infectious disease control and the rest of the world requires the health world to become versed in diplomacy. But it also requires the reverse: diplomats and ambassadors, attachés and envoys, getting in tune with the style and the importance, the humanity and value, of infectious disease control. Similarly, it requires the health world to tune in to new roles, smarter responsibilities: to opportunities to advance and resolve other global issues through dialogue and communications, involvement and negotiation, program design and delivery. Opportunities that become visible and tangible through epidemic response or other personnel internalizing a fundamental appreciation of diplomacy—the art of conducting negotiations between nations, in handling affairs without hostility [4]—in essence, the art of thinking globally, while acting locally.

By adopting such roles and making such contributions, global health can advance human rights: can address unacceptable or intolerable political, social, cultural acts by pursuing consensus-driven goals. At the very least, it can ensure that it is not exacerbating threats to world stability and cooperation: As JFK said, progress towards such goals cannot be achieved without concerted effort from all people, and all professions. Such effort is perhaps particularly required from the people and professionals who find themselves in times and places where bigger pictures are unfolding; in Lesotho or Zimbabwe or South Sudan, Iraq or Afghanistan or Sierra Leone. Because being apolitical doesn't mean being adiplomatic; retaining the apolitical privileges of medical interventions fails to justify turning a blind eye towards the world's other concerns (Fig. 1.5).

For anyone raised on the romance and intrigues of history and politics, on the importance of affairs of state and statesmanship, ententes and détentes, global health diplomacy is a welcome bridge. Barefoot efforts are intertwined with broader spheres of negotiation, influence, alliances, pacts, treaties, and

Figure 1.5 On a mountain airstrip in Lesotho, health professionals are often the only diplomats around. *Picture courtesy Sebastian Kevany.*

diplomacy: such *ad hoc* diplomacy can connect altruism with bigger pictures—embracing, appreciating, and addressing them through barefoot operators.

Those involved are rising to the challenge. Rising, appropriately, in an era where power and decisions, responsibility and influence, are becoming more and more diffuse: an era of crowd decisions, with choices lying less and less in the hands of the elite. Efforts at the advancement of world stability are now extending beyond militaries and academics and politicians—they now include infectious disease control. Each fractional contribution aggregates to make a difference: the programs and interventions, and the styles of their acolytes and practitioners, are being fractionally tweaked—without loss of primary goals—to advance global cooperation. This is global health as a multi-stakeholder public diplomacy, in the hands of us all: interactions between countries, including partnerships and alliances, are now driven by the barefoot as much as by the formal diplomat. The global health diplomats, who are acquiring new skills and responsibilities by blurring the line.

Barefoot diplomacy is therefore generating new standards of style for the 21st century: new systems are being developed for evaluating lives and professions based on their downstream impact—on their invisible effects.

The world is changing and with it, the change of gold standards: new models of style and valour for millennials and Generation Z to aspire to, which rely on the diplomatic and the altruistic, to save the day [5]. There has thus been an evolution of tastes, requiring new skills and styles which are based on mantras of effectiveness and sustainability, and cost effectiveness and visibility, in both personal and professional lives.

Broader roles are also increasingly necessary because global health is not altruism alone. Medical or other related workers, or any kind of international soldier [6] or surfer [7]—are barefoot ambassadors as well. These people carry with them on their travels associated responsibilities, whether desired or not: their home countries, lives, legacies are increasingly judged on that basis.

They are being judged on their ability to do no harm, on broader levels—by embracing local customs and humour; by repartee and conversation. They are learning the ability to tune into basic phrases and greetings, to know how to shake hands, or whether or not to take your shoes off before stepping on to a rug, or into a home: these are the ones who adapt to local working hours and styles without demur, and tune in to benign local attitudes and values [8].

Diplomacy can also be advanced in global health by developing an awareness of local economic conditions; by managing expectations of what is feasible for locals. It can be advanced through understanding that the rest of the world differs from the Western ideal, by being careful in the use of bling and tech: can be advanced by tuning into new economic, social, and cultural paradigms in which familiar rules and expectations no longer apply. And, as Stephen Murphy herewith demonstrates, can be advanced by an inherent sense of the need to combine one's duties.

In search of a revolution: a global health diplomacy journey

At university I majored in sociology, and dabbled in literature and philosophy. "Social justice" might best describe the subcategory of each discipline that interested me most. I'm tempted to say that I didn't have the most practical preparation for a career in global health diplomacy, but I wonder if I would have ended up where I have without this quixotic start. I wrote a final paper my senior year on "revolutionary pedagogy", and explored the links between Cuba's promotion of literacy education and how the Zapatista movement in Chiapas, Mexico established egalitarian social services in the jungle territory it controlled at the time. I had quite the appetite for studying revolutions—and even dreamed of teaching in someplace like Chiapas after college. But with

(Continued)

(Continued)

zero Spanish, zero savings, and only a few days of substitute teaching under my belt, the sober reality of adulthood took hold upon graduation.

Following that, I made one of the wisest decisions of my life and joined the Peace Corps. Any qualms over departing from my revolutionary path quickly subsided when the U.S. government agreed to teach me a foreign language, help me improve my teaching skills, and allow me to defer the significant student loan debt I had incurred. I taught English for two years in Cabo Verde, an island nation in West Africa with a strong democratic government and rising human development indicators. Cabo Verde also had an impressive revolutionary history, and I was fortunate to discover the writings of Amilcar Cabral, who gave his life in the struggle for the country's independence from Portugal.

While I prized the relationships I developed with my students and fellow teachers, I learned that planning lessons, correcting papers, and managing five classes a day of about 40 teenagers each was not my calling. I did, however, discover a passion for health promotion while mentoring teens for an after-school peer leadership program. I trained youth leaders on organizing peer outreach activities, focused on preventing HIV and teen pregnancy. With that, I stumbled into the field of public health, which eventually led to a career path of sorts in global health diplomacy.

When my Peace Corps Country Director asked if I would be willing to extend my service for a year to help start a Peace Corps program in Timor-Leste, I couldn't resist. It represented a chance to work in health promotion full-time, and an opportunity to live in a country that had just fought a revolution of its own. Although a quarter century of brutal rule by Indonesia came to an end in 1999, the United Nations ran a transitional administration governing Timor until May 20, 2002. I arrived just a few weeks later, during Timor's first month of independence. Indonesian soldiers, and the Timorese youth militias they exploited, destroyed nearly all the country's infrastructure as they departed in late 1999. The Timorese had won their freedom—but now faced extreme poverty, and the challenge of building a functional government, including a quality health system. During my year in Timor, I was inspired by the Timorese health care workers and government officials who led their communities in establishing health services. As a health promotion advisor, I planned vaccination campaigns with my Timorese counterparts, and helped incorporate health education activities into the daily work of the health centres in my town.

I also made the most of the presence of seemingly every UN agency and NGO under the sun, to network and learn how aid workers at these organizations charted their career paths in global health. Based on these discussions, I thought I might like to eventually join USAID's Foreign Service—to manage health programs, at the country level. I decided I should head back to the

(Continued)

(Continued)

United States after my Peace Corps service, find work at a health NGO to get project management experience, apply to grad school, and then try my luck at getting into USAID. I got most of the way toward meeting those goals before getting side-tracked by the diplomacy element of global health.

I joined Management Sciences for Health, a leading USAID implementing partner based in the Boston area, for a couple of years, and very much enjoyed my experience managing health projects in Africa. However, I started to have second thoughts about a career as a project manager. I became increasingly interested in the intersection of foreign policy and global health. I wanted a career straddling global health and diplomacy, and I didn't want to have to choose between the two for graduate school. The more I learned about the Fletcher School of Law and Diplomacy at Tufts University, the more it became the obvious choice for me with my eclectic interests in health, humanitarian aid, diplomacy, and management. I capped off my two-year program at Fletcher with an MPH in global health policy at Harvard, while serving as a Reynolds Social Entrepreneurship Fellow. While I was still quite interested in USAID, during grad school I successfully completed the hiring process to join the State Department's Foreign Service as a political officer. Knowing that the U.S. President's Emergency Plan for AIDS Relief (PEPFAR) was based at the State Department, I thought I might be able to find a sweet spot in global health diplomacy there.

As a U.S. diplomat I completed entry-level assignments in Brazil and Afghanistan. I then had a short-lived position in Sudan tracking the humanitarian situation in Darfur, until U.S. Embassy Khartoum was evacuated in September 2012 following violent riots in the wake of the attack on U.S. Consulate Benghazi (in neighbouring Libya). I made my way back to Washington, DC, and the prospect of returning to Sudan dimmed as the evacuation stretched from weeks to months. I went in search of a new position and serendipity struck. Secretary Hillary Clinton had just opened the new Office of Global health Diplomacy at the State Department.

Through this new initiative, the State Department sought to better leverage the United States' massive investments in global health to improve its standing with other countries, and to better leverage its vast diplomatic network to influence policy decisions that would improve health outcomes. I was fortunate to join this effort in its infancy, and help bridge the divide between U.S. diplomats and global health practitioners. I developed guidance for U.S. ambassadors and their political and economic officers, on engaging in health diplomacy. In developing country contexts, this often meant asking U.S. diplomats to influence policies that would improve access to, and quality of, health services: we encouraged governments to significantly increase their domestic resources for health, to strengthen global health security by investing in

(Continued)

(Continued)

disease preparedness and response, and to remove restrictions on human rights that serve as barriers to accessing health services.

In the Office of Global health Diplomacy, we also advised U.S. ambassadors and diplomats based in wealthy countries. We focused on using the U.S. government's influence to encourage these nations to contribute generously to the fundraising campaigns of multilateral organizations, particularly the Global Fund to Fight AIDS, Tuberculosis and Malaria and GAVI, the Vaccine Alliance. Through this work, I developed a strong interest in the effectiveness of multilateral partnerships.

When the Office of Global health Diplomacy merged with the larger PEPFAR office in early 2015, I joined the multilateral team at PEPFAR and managed the U.S. government's relationship with the Global Fund and UNAIDS. I advised Ambassador Deborah Birx and other senior officials on U.S. policy positions for the decisions coming before the board of each organization, and I served on the delegations representing the United States at board meetings. The highlight of my time at PEPAFR was negotiating the 2016 UN Political Declaration on HIV/AIDS, on behalf of the Obama Administration. In this agreement, UN Member States committed to a set of concrete actions to increase funding to fight AIDS—and to accelerate progress toward the UNAIDS 90-90-90 targets (90% of people living with HIV know their HIV status; 90% of people who know their HIV-positive status access treatment; and 90% of people on treatment have suppressed viral loads).

Following my Foreign Service assignment at PEPFAR, I joined the bureau at the State Department that manages relations with multilateral organizations, where I served as an advisor on humanitarian policy. While the position was a good fit, I soon came across an extraordinary opportunity to join the staff of the Global Fund. I decided to switch sides of the negotiating table, and liaise with the U.S. government on behalf of the Global Fund. The Global Fund raises vast sums of money from governments and the private sector ($4 billion per year on average) and provides grants to countries for evidence-based activities to prevent and treat HIV, TB, and malaria. The United States is the largest donor, providing roughly one-third of the organization's funding. I work with advocacy organizations to encourage the U.S. government to continue its leadership in defeating the three diseases. Diplomacy is a key element of my work. My colleagues and I work on a bilateral basis with donor governments to encourage them to fund Global Fund grants through our replenishment campaigns, and we also seek to influence multilateral efforts like the G7, G20, and various high-level UN meetings to advance the objectives of the broader Global Fund partnership.

For me, working at the Global Fund has been the ultimate global health diplomacy experience. Although I spend much of my time working on resource mobilization strategies in conference rooms in Geneva—and

(Continued)

(Continued)

occasionally speed walking the halls of Capitol Hill from one meeting to the next with Peter Sands, the Global Fund's Executive Director — I believe that I've finally found the right revolution for me. The Global Fund is spearheading a global movement that is succeeding at turning the tide against the biggest infectious disease killers of our time. This extraordinarily effective partnership saves millions of lives each year. When I consider what these lives saved mean to families and communities—and what they mean for the economies of developing countries—the word "revolutionary" truly does seem fitting.

Stephen Murphy

1.3 The barefoot diplomat

Diplomacy in global health can thus also be achieved through reductionism, through communications. Fractional, individual actions towards cooperation will aggregate towards greater world stability over time: the barefoot diplomat has to respect all levels, positions, castes and roles equally — locals and internationals have to be on the same page, wherever possible. Practical, day-to-day health diplomacy can therefore be achieved through involvement: through the dissemination of findings, and the facilitation of feedback loops. In this cotext, internationals cannot leave locals in ignorance; they need to involve and include them at every stage of the process. They need to ensure sustainability: that locals are not left wondering what happened to the HIV/AIDS clinic, to the malaria control program? Where did it disappear to; what was the impact on our public health? It is thus essential that internationals seek local opinions on how to improve program design and delivery.

Barefoot health diplomacy also requires qualities of neutrality and inclusiveness: often requires avoidance of political alignment; avoidance of taking sides. It perhaps has to recognize that neutrality may have to trump epidemiology in program location decisions; and that infectious disease control programs have to be considered in terms of the ethics and morality of their recipients [9]. Is the program inadvertently supporting the well-being of al-Shabaab, or Boko Haram? Are narrow idealists inadvertently providing health care to terrorists; is global health indirectly supporting extremists—or is, paradoxically, the lack of health care the true source of danger? If the latter, medical or public health emergency efforts may have an impact on resolving not only epidemiological but also security threats.

Barefoot health diplomacy simply means doing more: unifying antagonistic communities through non-discriminatory service provision, and by ensuring equal access and common clinic facilities across tribes or ideologies. Diplomatic global health programs encourage all to attend: they—whether in hospitals or clinics, caravans or tents—should be inclusive and accessible, even at the worst of times. They should perhaps be the *de facto* community centres: the places of refuge, of neutrality. Barefoot health diplomacy might best promote equal rights to treatment across races, ethnicities, social class.

Yet further uniting the aspirations of global health with other ideals will require negotiation and compromise: will require interpretation of achievement, or failure to achieve, and the weighing up of trade offs in broader contexts. It will require an ability to look beyond narrow targets, to negotiate with numbers, and to challenge frameworks (Fig. 1.6).

The barefoot diplomat has to negotiate internationally and locally but also between and across communities and villages; with doctors and head

Figure 1.6 Epidemic control community meetings in Northern Thailand. *Picture courtesy Sebastian Kevany.*

men, traditional healers and politicians. Barefoot diplomats have to navigate through all the diverse, competing, health and non-health agendas; between differing national, provincial and district priorities.

Such barefoot global health diplomacy involves cross-checking for opportunities through community involvement—via constantly considering human rights, and searching for opportunities beyond the humanity of health care for dual effects. Medical programs and practitioners have thus to re-examine, and ask if the work is in accordance with rights to free speech, to water, or to nutrition? Some programs and interventions are perhaps better than others at promoting human rights, while other programs deserve greater recognition for, say, the wider environmental effects they achieve.

In Thailand, I saw epidemic control negotiations work on multiple levels—in meetings that fostered involvement from everyone; all members of the community, all walks of life. From monks and merchants to sword dancers and shopkeepers; village elders to millennials: all describing their tastes, concerns and preferences; all helping to build the right, tailored infectious disease control program.

I saw also advancement of human rights in South Sudan, when the country was preparing for secession: the referendum looming on the horizon. The future was undecided—yet global health, inadvertently or intentionally, advanced the onset of the new political era. Of course, epidemic control initiatives there strengthened the health system—but they also helped to build the South Sudanese Ministry of Health, before there even was a formal capital city. In Sierra Leone, Nurse Amy Gildea discovered similar realm overlaps, in unexpected ways.

The guardian at the gate

I always envied those health professional—the wild and unconventional who pioneered and eked out a space for humanitarianism. The brave, fearless and possibly foolish, who charged forth into the great unknown in those early days with good intentions. Who honed both clinical skills, and a bit of a knack for *bricolage* or DIY.

I grew up with news clippings of Sudanese refugees on my walls as a teenager, and diligently learned and cultivated my French language skills so that one day I could be one of those pioneers. I wanted to be one of those "wild and unconventional people"—rugged individuals, who found connection in a larger purpose is how I had romanticised them in my teenage mind.

(Continued)

(Continued)

I carefully curated my path through university and various hospital roles to find myself a seat on that plane heading to anywhere, as fast as I could. That first trip delivered on all of those adolescent dreams: I arrived into Freetown in the early evening and took the helicopter flight across to the mainland full of excitement, and secure in the knowledge that a new colleague was there to pick me up.

That colleague never came.

Lesson 1: always have a backup plan, and a phone that works. My smile faded fast as night fell and I realised no one was there to pick me up — the reality and loneliness was as crushing as the humidity was on my lungs. The words of the supervisors in Paris rang in my ears — *"Someone will be there to pick you up when you arrive, but, if they're not, here is $2 for a taxi"*. That $2 was now burning a hole in my pocket: $2, and my phone with a useless SIM card that meant I couldn't call anyone. As the airport rapidly emptied out and I shifted my weight uncertainly from foot to foot, I surveyed the dwindling number of taxis waiting in the parking lot and knew I would have to make a choice.

A motley crew of Irish labourers noticed me standing on my own, and offered to give me a lift. This was my choice—this was my moment in the "choose your own adventure" where things could go ok, or very wrong. I chose right — those concrete puurers helped safely navigate me to my lodgings in a quiet backstreet on the outskirts of Freetown. I will forever be thankful. They also provided my first introduction to African musicians — the song *African Queen* would go on to be a defining song in the soundtrack of that period of my life. That unforgettable evening was topped off by a cold bucket shower, by candle-light.

Those months in post-war Sierra Leone were far from idyllic—and I certainly didn't feel like a hero, but it was an intense period full of learning. Sporadic weekend sojourns to the local river, or further afield to the beach, provided a welcome break from the long days and bleak conditions of the refugee camps. I quickly had to unlearn everything I had learned in my early nursing career, as a lot of it didn't make sense in this environment. We were overwhelmed with young children in the grips of malarial anaemia, obstructed guts, or other dire conditions.

Fast forward 15 years later — and, although I don't work so much on the frontline of global health anymore, what I have learnt is that the wild ones always need a guardian at the gate. Years and years of living in and alongside suffering can take its toll.

Lesson 2: finding ways to craft a more meaningful and extraordinary life must be balanced with a quietude that refills the soul. In addition to a more

(Continued)

(Continued)

balanced perspective, these days I'm a bit more streetwise than that young girl that landed in Freetown all those years ago. The world has also changed significantly — local SIM cards are always available in most airports; I carry a portable phone bank to ensure I always have battery power to call someone; and I usually have a bit more than $2 in my pocket!

RN Amy Gildea

1.4 Not stepping on toes

Yet barefoot global health diplomacy strives to avoid taking sides: tries to not promote one set of views, either consciously or unconsciously. It instead navigates the pressures, responsibilities, and awarenesses that such situations bring; thus creates demands on epidemic response workers, untrained in bigger pictures, who have likely only been schooled in epidemiology or anatomy. Such demands require all involved to have a broader *nous* and *savoir-faire*: otherwise, narrow-focused internationals will be all too easily swept into partisan agendas.

Barefoot global health diplomats are also required to be adaptable—personally and programmatically. They have to adapt, diplomatically, to unforeseen obstacles and consequences thrown up by the altruism of the health programs they are delivering; obstacles created by local conditions being different from what was expected or planned for.

Barefoot health diplomacy thus above all requires intelligence and an active mind. By practicing barefoot health diplomacy, there is also a solution to the *ennui* that can come with the endless waiting and downtime that accompany such a life. Global health workers—internationals and locals—perhaps need to use the dead hours to ask different questions; to look at situations from different angles. To consider smarter systems of service delivery; to think about better medical waste disposal as a wrapper blows past in the wind. In the field, you have time to work it out—plenty of time.

So, barefoot global health diplomacy involves attention to detail: beyond targets or job descriptions, integrating the quantitative and the qualitative. Paying attention to side-lines or afterthoughts, and fine-tuning the details that can make the difference between success and failure. Paying attention to use of local resources: is the program pirating local water pumps, or electricity, or diverting other essential resources? Such *ad hoc* diplomacy will thus leave only the footprints of better health, better relations.

The barefoot health diplomat also needs to consider local values, religions, and beliefs: needs to be aware of potential divides within communities, as well as divides both nationally and internationally: does the program challenge or undermine sensitive traditions; are the inherently benign qualities, as well as the malign, confronted? Does the intervention challenge habits deeply engrained in the culture, too quickly; does it impose alien, unpalatable beliefs upon recipient communities?

As individuals, also, we perhaps need to ask ourselves if aspects of personal styles, along with program design, can determine local acceptance and involvement with a program. The barefoot diplomat should perhaps thus ask if tuning in to local customs and practices through etiquette and protocol can make a difference: the individual in the field is the medium and catalyst for international collaborations; for partnerships forged through existing, albeit briefly, on an equal footing with locals; through the fostering of local education and advancement, when the chance arises. Examples of these partnerships can include gestures as simple as the transporting of a laptop computer to a local, or writing letters of support for their professional or educational dreams. By involving, also, locals in documenting achievements: through offering co-authorship of papers or reports, providing chances to voice opinions on the global stage. These are acts which go beyond job descriptions or requirements—yet, potentially, they are of inestimable value.

Professional partnerships, and partnerships of spirit, of *esprit de corps*, ultimately lead to the realization that local insights and advice complement, rather than threaten, those of the international. Barefoot diplomacy in global health hinges on the personal; on friendship and banter, repartee and camaraderie. It is dependent on making connections—and, under Murphy's Law, at times when one is likely least equipped and inclined to do so. Times when you are tired, or disoriented—in early mornings or after long flights, when too hot or too cold, when dusty or itching with mosquito bites—at

Figure 1.7 Nation-building is often indistinguishable from epidemic control. International support for ministry development in pre-secession Southern Sudan. *Picture courtesy Sebastian Kevany.*

times like that, you have to dig deep to advance diplomacy as well as health (Fig. 1.7).

And, finally, barefoot health diplomacy is dependent on honesty as much as charm. It is beyond making placatory remarks, just before departure: it means having awareness of short-term-ism; avoiding the temptation—because you are gone in a week, or two weeks—to take the path of least resistance. To not make promises of communication or evaluation that you can never keep; to not engage in equivocation and prevarication, before you get the plane. Successful barefoot diplomacy is, instead, dependent on keeping promises, not on raising false expectations.

This is a lot to consider: there is so much equipment required in the armoury of the barefoot health diplomat/maybe it can all be distilled into an awareness or a sixth sense of transcendent roles beyond targets and job descriptions. This awareness, often undefinable, is governed by instincts: it is an intuition of downstream effects, or consequences of actions. The many exhortations and standards, ideas and innovations, of barefoot diplomacy that follow can eventually become

Figure 1.8 Not quit bare feet — but close enough. Walking to site visits in rural Zimbabwe. *Picture courtesy Sebastian Kevany.*

second nature: the barefoot health diplomat knows, intuitively, that the price of peace is eternal vigilance (Fig. 1.8).

1.5 Key messages

- Conducting international relations and diplomacy efforts is implicitly within the purview of everyone involved in global health.
- This will become more explicit over time, and is already implicit in many efforts and endeavours — even if invisible to the individual.
- A balance needs to be struck between being aware of and fulfilling this role and attempting individual efforts in the diplomatic realm.
- The skills required to become a barefoot health diplomat are not many or complex, mainly depended on a cooperative approach.
- Barefoot global health diplomacy is likely to remain formally unrecognized for many, but has the potential to contribute to world stability (Fig. 1.9).

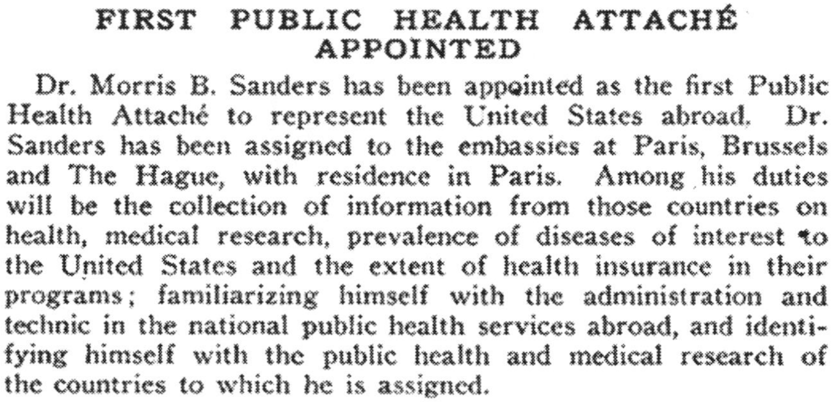

FIRST PUBLIC HEALTH ATTACHÉ APPOINTED

Dr. Morris B. Sanders has been appointed as the first Public Health Attaché to represent the United States abroad. Dr. Sanders has been assigned to the embassies at Paris, Brussels and The Hague, with residence in Paris. Among his duties will be the collection of information from those countries on health, medical research, prevalence of diseases of interest to the United States and the extent of health insurance in their programs; familiarizing himself with the administration and technic in the national public health services abroad, and identifying himself with the public health and medical research of the countries to which he is assigned.

Figure 1.9 Health and diplomacy have a long, but often unrecognized, historical synergy. *Courtesy of Matthew Brown.*

References

[1] Kennedy, J. *Address to the United Nations General Assembly.* September 20, 1963.
[2] Novotny T, Adams V. Global health diplomacy—A call for a new field of teaching and research. San Francisco Med 2007;80(3):22—3.
[3] A further development of the McNamara Fallacy (see "Deep Waters").
[4] Oxford English Dictionary definition of diplomacy (2017).
[5] Kevany S. James Bond and Global health Diplomacy. Int J Health Policy Manag 2015;4(x):1—4 Editorial commentary.
[6] Kevany S, Baker M. Applying Smart Power via Global health Engagement. Joint Forces Quarterly 2016;83 4th Quarter.
[7] Kevany S. Act Locally. Surfer's J 2016;25:4.
[8] Kevany S, Khumalo-Sakutukwa G, Murima O, Chingono A, Modiba P, Gray G, et al. Health diplomacy and adapting global health interventions to local needs in sub-Saharan Africa and Thailand: evaluating findings from Project Accept. BMC Public Health 2012;12(459):1—11.
[9] Kevany S. Global health engagement in diplomacy, intelligence and counterterrorism: a system of standards. J Policing Intell Counter Terrorism 2016;11(1):84—92.

CHAPTER TWO

Adaptation: epidemic control and local style

Abstract

Global health programs are often designed, in their original forms, for use in developed country settings. Extensive efforts are now made to adapt programs to developing world environments, with a focus on affordability and functionality in resource-poor settings. However, there are other dimensions of adaptations to global health program design and delivery, as well as epidemic control and public health emergency responses, that should also be considered—most notably from the diplomatic perspective. To what extent does a health program—even when adapted clinically—respond and fit in to local culture, religion, economics—to social structures, and sensitivities? From the choice of which intervention to use to the branding and logos that are required, the diplomatic adaptation of an intervention can have a key impact on service utilization and acceptability in recipient communities.

2.1 Adapting the global to the local

Global health succeeds only if it brings dignity, worth, and health to local people. It succeeds because epidemic control adaptation is ultimately about diplomacy—about finding a mutually acceptable way to give locals what they want, often through building bridges and partnerships (Fig. 2.1). Such programs often do a lot more than meets the eye, even if their effects are often not captured, measured or recognized [1].

Yet global health has to adapt if it is to compel and win over not just the locals, but also the internationals: not just the recipients, but also the donors. It should also perhaps adapt because, if the North and the West are to continue to support improving the health of poor people in poor places, it is because of the cool projects, the smart projects—the efforts that stand out, those that have the 'X-Factor'. Programs and responses and interventions that adapt to address issues not only of health—but also recognize conflict, the environment, inequality and poverty. The programs that end up helping the world as much as the locals: the programs that demonstrate that to respond to epidemics or emergencies or strengthen related health

Barefoot Global Health Diplomacy.
DOI: https://doi.org/10.1016/B978-0-12-818681-7.00003-0

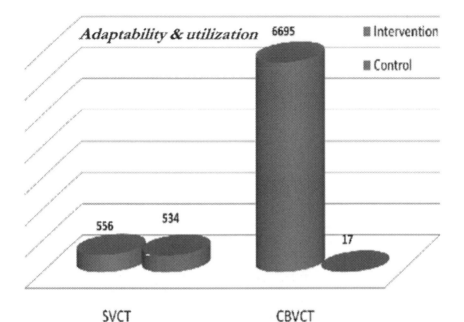

Figure 2.1 In many epidemic control programs, adaptable programs can make critical differences to service utilization. *Picture from Khumalo-Sakutukwa, G., Morin, S. F., Fritz, K., Charlebois, E. D., Van Rooyen, H., Chingono, A., ... & Sweat, M. (2008). Project Accept (HPTN 043): a community-based intervention to reduce HIV incidence in populations at risk for HIV in sub-Saharan Africa and Thailand. Journal of acquired immune deficiency syndromes (1999), 49(4), 422.*

systems, either directly or indirectly, is also to support other global goods [2].

But, what does it look like in practice? Whilst working in Kenya some years ago, I happened to see addenda to progress reports describing *ad hoc*, hidden achievements which were accomplished through adaptation. But these downstream benefits slipped through the cracks, because only the functionalism and the narrow effectiveness of the program survived in the reports: the international bosses missed descriptions of unexpected diplomatic outcomes of global health efforts: of community reconciliation, conflict resolution. A typical afternoon in a boring office in an exciting place involved reading the addenda, and stopping to think: this could be big. This could be what everyone wants, but no one sees- the ultimate reason for designing and funding smarter global health programs.

Maybe I was lucky to see it. Lucky to have had the eyes to see it; to have been schooled in economics and politics before public health. Lucky to be seeing the same things as everyone else engaged in the programs, but with a slightly more off-beat perspective. Seeing that adaptable HIV/ AIDS, tuberculosis, malaria or other epidemic disease programs work magic beyond health: at the same time, I was also seeing and questioning the rigid systems of program design; seeing how to adapt infectious disease control efforts to respond to different values and agendas. These values— strategic, tactical, diplomatic—advanced by downstream effects; by medical programs responsive to both local and international health and non-health needs.

I was thus, in that airtight office, perhaps seeing that global health can be good for everyone—a win-win situation—if adapted to enhance stability, security and trade. It was something that could advance global interests and address global concerns, while giving poor communities what they want and need as well—not to mentioning containing the global spread of disease. I began envisioning related efforts evolving diplomatically: envisioning programs demonstrating humility, inclusiveness, flexibility, and a will ingness to learn; programs that adhered to local priorities, respected local opinions, appreciated local style—thereby simultaneously advancing health, environmentalism, conflict resolution, stability, and diplomacy. I began picturing systems that were adaptable to circumstantial demands beyond protocols; that could sidestep siloed mentalities (Fig. 2.2).

But I envisioned global health programs changing shape for diplomacy mainly because we live in a blurred, overlapping, wired world. No environmentalist wants infectious disease control programs to leave medical waste—used products, empty packaging left behind. No-one wants litter left by the internationals blowing in the wind across the village square or the countryside, as the only legacy of the aid circus having come to town. Global health in the 21st century, should perhaps align with environmentalism just as with economics, security, politics, culture and society.

To advance one set of values of improving the health of the world's poorest people—utilitarians won't mind if the environment suffers in the process, as long as they fulfil their own institutional agendas—maybe appropriate to a world where everything is compartmentalized. But not to the global world; not to a world where everyone sees everything; in which everything and everyone is connected. These values are perhaps no longer appropriate to a world in which everything— every act, program, mission, decision, intervention, profession — bleeds into everything else.

Figure 2.2 In Kenya, even the working hours had to be adaptable. Searching for a missing tire en route to a nocturnal site visit. *Picture courtesy Sebastian Kevany.*

Over time, I began seeing the need to adapt so that epidemic control efforts jived with bigger pictures. On the hundredth conversation with a local nurse in the hundredth run-down rural health clinic on the hundredth mission, I was slowly gaining insight and awareness: understanding that HIV/AIDS, tuberculosis, malaria or other epidemic disease programs also had to strengthen the independence and resilience of the country; appreciating how global health also has a duty to moderate extremism, or continue to build trust with distant donors.

I was finally seeing the benefit of the medical adapting to culture, gender, religion, environment, epidemiology, economic. A new combination, because of new realities - the realities that global health, its programs, its acolytes and protagonists, increasingly have to hone instincts across wider fields of vision. Adaptable programs, with an ear to the tracks, can thus generate opportunities to do remarkable things for the world - within and beyond health. In similar ways, our next essayist, Dr. Annamarie Sehovic, found that unless messages could be understood by local patients in South Africa, a lack of communication would inevitably constrain aspirations.

Translating health; health as language; health as dignity; health as life beyond living

Circa October 2003. Lawley, squatter camp on the outskirts of Soweto.

In Lawley, a squatter camp outside of Soweto, the cuffs of my pants turned crimson. A chimney the same colour stood incongruously in the middle of a dusty plane strewn with refuse: paper and plastic bags; half-eaten potato chips; empty jugs that once held water, or milk. Two such jugs stood neatly on the blue counter in an impossibly small house, whose ceilings I could touch.

"Danny" and I were in Lawley to talk about nutrition in the face of HIV. His Anglicized name was a misnomer, a mistranslation; a simplification for those who could not — or would not — pronounce Sibongile. We went into a brick room which served as a church. It was open to the breeze and dust and birds; exposed to the elements outside, exposed to the elements inside. I was there to explain what the human immunodeficiency virus did inside the body — and to try to highlight the importance of adequate nutrition in order to stave off the starving effects of an ailment and its sufferers then referred to as 'Slim'.

Sibongile was there to translate, into Sotho. At the end of my talk — where HIV comes from; how it works; what it does to the body of the infected; how it affects and infects further — a man stood up at the back: *"Could you please say everything again, in Zulu".* I talked. Sibongile translated.

Every week more people came. Women. Then men. Every week, Sibongile and I tried to explain, to translate; to highlight the role of nutrition in staving off the effects of the ravages of HIV:

In order to fill the jugs in the kitchen, their owners had to walk to water.
In order to get food, people had to walk to markets, to small stores.
In order to get clothing, people had to walk to stalls in markets.
In order to get medicines, people had to walk.

But the HIV-anti-retroviral rollout in South Africa was still years away. Every week, Sibongile and I came and talked and tried to promote a soy-based nutritional supplement meant to support life — SoyaLife — in the meantime. The meantime could be short. I was invited into the chimney once: the man inside lay on a pile of sheets, regularly shifted and changed by caring neighbors. Neighbors left food outside the chimney, careful not to come in — to a space that would barely admit two people standing, let alone lying down. The chimney was narrow and high: it reached the sky, which shone through.

Translating health relies on language. Literal language, body language. Translating health is also translating dignity: the bestowing of a name; the caretaking of life beyond merely surviving, beyond just living.

The act of translating health — into literacy guidelines, into public policies, into international architectures and global strategies — appears to fall into the remit of professional diplomats. In the recent past, health diplomacy was

(Continued)

(Continued)

relegated to development aid workers—delegated to doctors and epidemiologists, employed by various centres for disease control and prevention. Today, more and more countries are building whole cadres of health diplomats.

These professionals serve as staff at national embassies, and at the international level at the World Health Organization (WHO). They are present at the Foreign Policy and Global health Initiative, at the United Nations. Some of them are drawn from the ranks of national ministries of health, or health agencies. Few of them know what it means to be health illiterate; a number of them likely know what it feels like to have the pronunciation of their names mangled. Not many of them have, in all likelihood, experienced the attempt at living in a roofless chimney. None of them have probably ever contemplated dying in such a space.

Yet health, especially global health, depends upon the translation of localities such as illustrated above, into global politics and policies. Diplomats represent a key bridge in this process. However, given the gaps outlined, how can it be possible for such professionals to translate health from such localities to the global reaches of power?

These professionals need to be barefoot diplomats—literate in the languages, cultures and codes both in local conditions and in the global corridors of agenda setting and decision making power. They have to evince humility and conviction: such health professionals embody these ideals. They are part of the reality imperative to realizing local and global health by successfully translating the languages of health, by reflecting the dignity of human life, and by facilitating life beyond living.

Dr. Annamarie Bindenagel Sehovic

2.2 Listening to locals

Adaptability, in life, sometimes only works in theory: sometimes the need to be adaptable, whether in controlled environments or special circumstances, is only evident in retrospect. The same truths apply to crafting smarter global health efforts: there is the same need to bridge the practical—the day-to-day which is on the ground at intervention, program, personal, organizational, field levels—with the philosophical, the abstract. Throughout my decade and change of around-the-world missions, I was finally seeing the need for the program that was adaptable to circumstances: weather, distances, heat, cultures, places, war, ethnic tensions, geography. I was seeing the need for programs that could change shape; to jive with local styles but also to make magical, domino effects happen. Adaptability is thus necessary for interventions to be appropriate and responsive: just like the Darwinian parallel, if such programs are indifferent to their environments, they will struggle to survive [3].

Adaptation is defined as the ability to change, be changed, or to fit changed circumstances — according to the dictionary, "to make suitable to requirements to conditions; to adjust or modify fittingly" [4]. It is sometimes a conscious decision, sometimes an unconscious modification: the individual is adjusting to cultural, security, social, or economic surroundings without really thinking about it. The global health program — the epidemic or public health emergency response—like the individual, needs to fit in: to expound flexibility, to evolve as far as it can; to thrive, while also staying true to itself.

Adaptability is relevant to all forms of global health: to the trials, programs, and international efforts; to the bilateral, to the multilateral. But all epidemic control and response, in all its forms, is governed by the didactic, the technical; by protocols and operating procedures. The same structures apply to any area of life, any job or profession: rules, goals, outcomes and targets. Procedures and frameworks for setting, measuring and achieving these goals and outcomes are governed by internationals and theorists often located thousands of miles away from the work that is taking place: in high office blocks in modern cities, far from the dusty roads, and the wood smoke.

Global health is thus often a clash of cultures. For the office-bound workers, it's about devotion and fidelity to the protocol: it's about maintaining respectable, ethical connections to plans and designs; to what you promised to do. You can't promise to deliver voluntary counselling and testing programs for HIV/AIDS and instead, treat malaria; build a clinic when you've promised a hospital. From the office perspective, the work is about the need for connections, in fundamental and essential ways, to original plans: it is about adhering to core principles and practices governed by parameters and limitations, and by ethical and operational boundaries that determine what can and can't go down. In voluntary counselling and testing for HIV/AIDS, for example, there is a need for recruitment guidelines about age of consent; the right to confidential results; to disclosure and privacy. Protocols and standard operating procedures often thus protect patients and staff; but they also determine service opening and closing times, along with standards of care and safety, communications and security.

Fidelity to protocols and procedures is essential, and so are rules—but there's always a balance. On the ground, and away from the office, adaptability is equally essential. From afar, it is often overlooked, not always encouraged—but it has to be part of the consideration of what is happening in the field. The medical has had to become more and more be about considering local needs, expected or unexpected; providing what locals actually require, as well as what they are told they need. This is where adaptability comes in: it's about working out what is going right, and what needs to be changed, for emergencies in health and beyond; finessing structured rigidity to try to do something cooler and smarter, when one can. Creating something better, more multidimensional, is thus part of the art, the beauty, the creativity of global health [5].

Adaptability is also about anticipating the unexpected; anticipating the reality, as much as possible, when theorizing. Inevitably, you can only know so much *ex ante* about local demands, proclivities, preferences, styles, needs and conditions. You can't know everything, before you get on the ground: before you get off the plane, out of the taxi, out of the hotel. You have to learn that flexibility over time, as well as at the start of a project; it is essential to making global health work. It's essential to stay *à la mode*, up-to-date, even fashionable—in touch with the day-to-day. Malleability, as much as rigidity, will determine success. Flexibility and responsiveness—shape shifting, if you will—are what bridge the gap between the theoretical and the practical; between efficacy and effectiveness; between the cultures of office and field (Fig. 2.3).

In the international medical realm, adaptability is also needed so that international programs dovetail with local requirements and involvement. What is needed, therefore, are adaptable global health programs that are planned, structured, designed and delivered by locals and internationals together—equally. The prescriptive, the didactic, the North-to-South approach to infectious disease and epidemics has resulted in the disadvantage of low input by poor people, and poor countries, in programs that affect them. The idea that better education implies better understanding, better knowledge, greater practicality isn't always true, and maybe never was. Most of the time, it is the locals who are going to come up with the barefoot solutions, and the renaissance ideas.

It's only the locals who can tell you to change responses for weather, for schedules, religions, cultures or conflict. It is the locals who provide insight to and an awareness of sensitive situations; who see invisible social or cultural or behavioural boundaries, and know how far you can stray beyond them. In terms of what global health says and does and promotes; in terms of permissions and approvals; in terms of how, sometimes, to be ethical, you have

Figure 2.3 In rural Zimbabwe, when community centres were unavailable due to political or other constraints, participants made do elsewhere. *Picture courtesy Sebastian Kevany.*

to bend the rigid parts of a program to make it work. In Zimbabwe, during my time there, it was the locals who negotiated with local politicians to ensure that nuanced HIV/AIDS programs could be delivered — even when the official line was that a different program was to be implemented.

Locals thus prove that adaptation in epidemic response programs can be a catalyst, a spark for bigger changes; a window to new kinds of double planning and multilevel awareness. Local people will adapt programs to stop disease—but also to sort out whatever else they can as well. The internationals hence need to listen to the locals—either through casual exchanges at community working groups, or on formal occasions with advisory boards in the middle of a field or in an empty schoolhouse. Listening in a village street, to an old man or a woman with a baby on her back: listening to what they felt threatened by, what they disapprove of, what they want more of. Listen to ideas about what else could be achieved; harmonizing global health with the economic, the social, the environmental, the local [6].

Of course, structure is required for locals to have a voice to take on governance: Structures for voices of field staff, villagers, patients or anyone with ideas that would make things work smoother, cooler, smarter and better to be heard. Structures to give voices to innovative ideas that would otherwise remain unheard: that allow those coming up with new ideas in crowded waiting rooms in rural Zimbabwe to write them down; structures for sending those ideas to Harare, and on to the internationals. Structures for locals and internationals to make decisions together; to fine tune those ideas in collaboration; and to send back quick approvals—and back out to the rural edges.

Approvals for change are received; then, suddenly, change is happening. Suddenly, in far-flung Mutoko, HIV/AIDS epidemic control efforts are overlapping with health education, community empowerment and income generation. Suddenly, schedules, services and styles are changing to fit with the local groove. Yet this isn't always the case: communications and turf battles or other inflexibilities can nail progress of even the most benign concept, as our next essayist, Dr. Barry Levine, illustrates

Challenges in deploying medical record system in developing countries – a personal experience

In recent years, we have gained valuable experience in deploying medical record systems in developing countries. The challenges come in several varieties, including via governments, clinics, hospitals, local staff, and the lack of on-site expertise and infrastructure. Our global collective experience by now includes sites in Africa, Asia, South America, and the Caribbean; we have deployed systems in clinics and hospitals, as well as providing telemedicine services. The following notes are drawn from the author's experience in deploying an open source Electronic Medical Record (EMR) system in under-resourced settings.

Regarding government interest in deploying an EMR system, it is important to carefully consider the national health impact—as well as the political impact, and legal issues. An example encountered by the author was the following: the government in question indicated, due to limited available funds, that their preference was to budget for tangible goods, e.g. medicine, rather than budgeting for an EMR. In fact, political officials were not interested in an EMR more due to increased transparency provided by the system, which would impact corrupt practices!

It is also worthwhile to consider the location of deployment of the EMR. If the EMR is on a server maintained locally, then there must be staff to oversee the operation of the hardware and network. The staff might include computer engineers, who would be involved in maintaining the hardware, as well as working with non-local computer expertise responsible for maintaining the EMR system.

(Continued)

(Continued)

This system maintenance might include hardware maintenance, diagnosing software errors, or system upgrades and the development of new localized software.

If the EMR system is deployed in the "cloud," then one must consider associated benefits and consequences. The consequences include the charges incurred from the provision of such services, as well as associated security issues and necessary local infrastructure — power, Internet availability. An associated legal aspect concerns the location of patient data: since the information would no longer be stored within the country, legal issues with respect to patient data rights would no longer apply. However, in this scenario, there is no need for local staff for hardware maintenance. In addition, system upgrades and software oversight can be performed at a distance. Difficult choices. . .

Whenever a new system deployment of the EMR system arises, we also consider the necessity of a local champion. When moving from a paper medical record system to an EMR system, there will always be an impact on local staff. Locals need to be trained in the use of the system—as well as the implication of workflow changes. We have often observed a project failure wherein the project manager did not pay sufficient attention to the impact on local stakeholders at the clinic. These stakeholders were very concerned about the impact on their workloads; the project manager was not able to provide ownership of the project to the stakeholders. Therefore, the stakeholders were not willing to cooperate to ensure success of the project.

All too often, local stakeholders are not adequately engaged in the change process—so they do not cooperate to ensure success of the move to an EMR. In short, a local champion has to be identified and assume the responsibility of ensuring success — local ownership, if you will. An anecdotal experience the author recently encountered involved a clinic leader, who kept wondering why the author had volunteered to deploy the EMR system. Despite a positive meeting between the author and the country's prior head of parliament, the clinic leader was still unsure of the author's motivation. It is likely the deployment, in that case, will fail.

In summary, the key points to ensure success of EMR deployment include:
- Identification of a local champion who will be responsible for overseeing the project with the local staff;
- Obtaining a good understanding of local infrastructure;
- Providing ownership of the project to the local champion, as well as the local staff;
- Ensuring the local champion will commit time and local resources to the project; and
- Ensuring the local champion and staff maintain an open line of communication.

Dr. Barry Levine

2.3 A thousand ways to adapt...

On the ground, the success of global health adaptations depends on the micro-level: whether the opening and closing times of clinics on public holidays in Zimbabwe; or the phraseology of health-education messages in Kenya. Yet it also depends equally on the big picture, on the macro-level: on international exchanges becoming less didactic, paternal, adversarial and more inclusive, collaborative, equal; on the internationals flowing with local needs, but in alignment with global prerogatives. On the locals and the internationals tuning in to the strategic, the diplomatic, so that national values—perspectives and viewpoints, traditions and aspirations—don't get overlooked, or brushed aside. Adapt like that, and it's a win-win [7].

Adapting to local needs, considering local views and taking local advice is critical to achieving the big picture in global health terms: to responding to tuberculosis, malaria, Ebola, Zika; to containing HIV/AIDS epidemics. Adaptability is thus critical to the success of internationals asking locals to change day-to-day principles, beliefs, habits, practices—for their own health, and for the health of others. The internationals, as well, may find that there is nothing harder than changing habits; that it is difficult and upsetting, that the pursuit of the big picture requires sensitivity and care in program design. Requires, also, flexible programs that make sure that whatever epidemic control or other public health efforts asks of locals is manageable, clear, practical and acceptable.

Above all, adaptation is necessary for health. In Zimbabwe, our team was providing voluntary counselling and testing for HIV/AIDS amidst a cholera outbreak. Locals infected with cholera were coming to the nurses in the testing caravans and tents, saying: we are in trouble—can you help us, because no one else is around? This caused ethical dilemmas. What should the nurses do—stay rigid or flexible: don't change anything, or do what they can? Fortunately, the decision was made to evolve; to change the shape of the program partly for health, partly for diplomacy—but, above all, to give the locals what they needed. The service offered and the treatment provided was quickly changed to respond to the emergency; infectious disease control is as much about the day-to-day as the protocol.

Epidemiological adaptation is necessary as well—to respond to nuances in epidemics, or rare forms of disease. Awareness of local mutations, of the evolution of viruses, and of generalized and concentrated epidemics can be vital. Asking about which population groups are affected, and if some are at greater risk than others, can make all the difference. Of course, these answers are not always clear. Is it effective to focus on sex-workers' rights for HIV/AIDS treatment; is it inappropriate to jointly test married couples; or, are they—ironically, paradoxically—exactly the ones who need the help? Epidemiology thus forces changes to design, making programs more responsive to get to those in need—to the populations, regions, districts, villages, locals, that are most vulnerable. Epidemiological adaptation therefore requires responsiveness to vulnerability, science, new findings and discoveries: requires embracing better systems of infection testing, or screening for tuberculosis. Reacting to new findings requires flexibility—because if the world, the circumstances, the environment and even the disease are always moving forward, always changing, then global health has to as well [8].

Adaptability to function can be so important, as well. There can be shiny new clinics, tents or caravans in the middle of a field all day—yet no one coming to avail of the program; the nurses getting restless, bored and demoralized. Global health successes are thus dependent on linking to culture — to, for example, local gathering places, rituals and routines.

In Tanzania we delivered HIV/AIDS programs at water pumps. The water pumps were where locals would gather amidst work and family, amidst stacks of brightly coloured buckets: there they were accessible. There they welcomed us, spoke with us, listened and took an interest: an exercise in adapting location to culture; suddenly, everyone was coming; everything was clicking.

In Tanzania as well, we culturally evolved programs by changing logos, names or terms to better suit local tastes—by changing English to Swahili, 'Project Accept' to *afiki*. We evolved by involving traditional healers; by showing respect, asking advice; and by including—instead of impinging, threatening, marginalizing. We adapted culturally by turning up when invited: in the middle of the night, to tribal rituals and sword dances, to bonfires, concerts, vigils, and mystical full-moon rituals; by turning up in the right way, at the right time—leveraging culture, without exploiting it. Cultural adaptation can often be a part of the global health response mosaic so easily overlooked by distant planners (Fig. 2.4).

Figure 2.4 Conducting epidemic health education sessions at the local water pump in villages outside Kisarawe, Tanzania. *Picture courtesy Sebastian Kevany.*

Cultural adaptation can lead to social adaptation; to health services for males, females, or both: together or separate, integrated or segregated. In Afghanistan, a malaria program had to respond to unexpected local demands for segregation; local women were declining integrated services, because suspicion and stigma were keeping the vulnerable away. Adaptable epidemic response efforts sometimes has to sacrifice *beaux ideals* of social progressiveness for *realpolitik*—it's a never-ending balancing act.

Global health thus has to be adaptable to culture, to society—and to religion. Back in Tanzania, we found ourselves trying to concoct programs that recognized diverse religious beliefs—Christian, Buddhist, Animist, Muslim, Hindu—all in one village. Over in Thailand, we were adapting to *avant-garde* combinations of Buddhism and Christianity, where, in coterminous villages in far out border jungles, locals identifying with one, the other—or both. There were ostensibly distinct village cultures, yet they were blurred: our program had to handle an often surreal interchangeability and integration of beliefs, as necessary. In Zimbabwe, meanwhile, we were adapting to neo-Afro-Christianity; villages of bibles, piety and gospel quotation. There were similar situations in Southern Sudan—adapting to the uncategorizable; and in Afghanistan, Jordan, Iraq, Egypt, to the ever-changing face of Islam.

Adapting to religion in each place led to tricky questions—such as to provide, or not provide, health services at churches, mosques or temples? Should one use identical or different styles for Buddhists as for Christians? Was it unfair not to, or folly to attempt consistency in diversity; was the mere idea of dovetailing public health and religion sacrilegious, in some way? Even if our programs were approved by the priests—the shamans, the clerics, the witch doctors or the monks—what about the rest of the locals? What about the villagers, and their beliefs?

Even with approval from local religious leaders, we found ourselves adapting again for appropriateness. In Zimbabwe, when locals walked past, solemnly, in their best clothes, on their way to church,—eyeing us suspiciously—we felt the needed to change. Felt the need to respond, as the sound of Swahili gospel song echoed across the bush: realized there is a time and place for everything, even with community leader approvals. We had thus also to evolve and adapt in line with local interpretations of tradition and rules; had to react to locals in distant villages regarding the place of belief in their lives, in their hopes, in their survival—particularly when our efforts meant traditions were challenged.

Over in Thailand, finally, we got the balance right: We sat with Buddhist monks and villagers in the highlands near the Cambodian and Burmese borders. Sat cross-legged, on bare floors, seeking enlightenment—advice, approval, appropriateness. The monks and the villagers approved holding field-days for infectious disease risk education at their temples.

Approved it, but making it appropriate in ways that flowed with local styles: an example of villagers and monks guarding against the rapacious, the utilitarian; against private-sector mentalities overlaid on global health; against leveraging solemn ceremony as a measure to improve numbers, reach targets. The villagers and monks, together, were thus ensuring the non-invasiveness of our efforts: this was community led, community owned. Infection testing outside temples was allowed—but only after it was adapted and vetted through local eyes first. Our next essayist, Mabvuto Mndau, illustrates that epidemic control efforts, even I terms of financing, have to mould themselves to local conditions (Fig. 2.5).

Figure 2.5 In Thailand, no international epidemic control effort stood a chance of success without Buddhist approval and input. *Picture courtesy Sebastian Kevany.*

Implementing results based financing and diplomacy: experience from Malawi

In 2009, the governments of Norway and Germany jointly proposed a contribution of US$ 10 million to Malawi for reducing maternal and neonatal mortality with a focus on safe motherhood and new-born care using results-based financing (RBF) mechanisms from October 2011 to October 2014. The programme was later financed by Government of Germany from November 2014 to May 2018, and implemented in 33 health facilities in four districts of Balaka, Dedza, Mchinji and Ntcheu.

The objectives for using results-based financing were two-fold: both to increase the quantity of institutional deliveries, and also improve the quality of care at participating facilities. The RBF approach had three components: investments in minor infrastructure works and essential equipment to bring selected maternity services up to a minimum standard to participate in the RBF initiative; introduction of quality and performance contracts (called performance agreements) for the qualified public and the Christian Health Association of Malawi (CHAM) facilities offering maternity services; and the

(Continued)

(Continued)

conditional cash transfer (CCT) for women to partially contribute to the out-of-pocket expenditure associated with travelling to and from the facility (and for staying in the health facilities for 48 hours post-delivery).

The overall programme responsibility lay with the Ministry of Health (MoH), and the Reproductive Health Directorate (RHD) was in charge of implementation, Consultants provide technical and financial management: the RBF approach was new to Malawi as the country was used to input based financing. To ensure clarity, understanding, integration and acceptance of the programme, there was need for continued diplomacy at national, district and community levels.

The programme ensured sustained partnerships, with old and newly-appointed government officials that came into office due to high staff turnover at national and district levels. At national level continued discussions, negotiations, meetings and field visits were held with law makers (members of parliament), as well as policy makers at Ministry of Health and Population. This approach helped to create support at national level, thereby easing programme implementation, policy development and policy implementation. The involvement of different directors in the Ministry of Health also ensured proper clarity and adoption the RBF strategy. Consequently, the strategy was incorporated in the health policy as one of the health financing mechanisms that could improve access to quality health services in Malawi. At the national level, the Principal Secretary and all directors at the Ministry of Health took active roles in guiding and supervision of the programme.

At district level, the diplomatic meetings involved district councils, community leadership and health workers. At council level, the diplomacy involved District Commissioners, civil society organizations and elected political party leaders. This ensured buy-in and advocacy of the programme at district levels. Additionally, at district levels, there was devolution of where new members were elected in the course of programme implementation following the country's tripartite elections in May 2014: the newly elected District Council members needed to understand the programme so that right decisions about the programme were taken. Before the elections of the councillors, the district commissioners were in command of the all health and development decisions. However, all council members had not been exposed to the RBF approach; this required continued diplomacy and capacity building. The capacity building also involved new members of parliament, chiefs and ward councillors in all the four districts so that their roles were well understood. These efforts were continuous to ensure a good level of knowledge, understanding and support at the district level.

At health facility level, the health workers did not have prior knowledge about the RBF programme, as the sector was used to input-based financing. The health workers were not always in compliance with standard operating procedures, thereby contributing to low quality of health services that were rendered to clients.

(Continued)

(Continued)

It was challenging for health workers to establish, maintain and follow standard operation procedures: the programme ensured continued diplomacy and support to health workers to ensure adherence to standard operating procedures, thereby improving quality of services delivered to the clients and general management.

At community levels, the conditional cash transfer component was initially misunderstood. There were myths and stories, and it was observed that a significant proportion of women who were registered and verified to receive the support did not receive the cash. There were stories circulating which suggested the cash was a way to attract women to facilities so that their blood and placentas could be collected, and used for nefarious practices. This had the potential to scare the women away from delivering in the health facilities. The programme thus developed diplomatic approaches, involving community gatekeepers and district councils, to engage community members about the programme. Health diplomacy at all levels was vital for improving access to quality health care.

Mabvuto Mndau

2.4 ...And a thousand more (ways to adapt)

Around the world, the locals act as the catalysts, tuning you in to the environment; but environment as culture, religion, society as well as the actual, the physical [9]. Locals see its importance, dominance and irrefutability: in Ethiopia, I was working on antiretroviral adherence programs far outside Addis Ababa, in the highlands around Gondar and Bahir Dar. Out there, there's no tech—no hardware, no equipment—so the programs had to adapt to that bare-bones environment by changing resource mix inputs appropriately. The visitng internationals, seeing the changes working, gained an appreciation, respect, and awareness for the physical environment and its limitations and opportunities. We came to appreciate improvisation: learned to sidestep the knee-jerk of importing, spending and buying on Western lines. Instead, we learned from the locals, looking what was there, and what could be locally-used to solve the problem.

The physical environment also includes the weather and climate, and its daily or seasonal tricks and routines. In each place, one has to try to decode it: to solve the riddle of what is going to work, when and where. The weather engenders mindfulness of the time of year:you develop awareness of rainy seasons and flooding that cuts off access; awareness of local routines that change in the sun and rain; awareness of flood plains, harvests, and inaccessible villages.

Global health programs thus need to be able to change according to weather and climate—both physical and political. Infectious disease efforts have to be ready, even if reluctantly, to adapt according to the realities of local and international politics: in Zimbabwe, in a highly-strung political *milieu* where public meetings were frowned upon, it was necessary to toe the line: to make distinctions between community outreach and political activism, to avoid suspicion of straying from health into the rebellious, the revolutionary [1].

Adapting global health to politics, in situations where unexpected gatherings of locals learning about infectious diseases can cause suspicion of inciting revolt or unrest, means taking care to advance ideals without ideologues. Epidemic response efforts have to adapt to situations in which the army and political officers are looking on—asking if you are starting a revolution; plotting a *coup d'état?* Epidemic control efforts have to evolve to deal with *realpolitik*—because, as discussed, neither the medical nor the world can compartmentalize as much as it once did. Infectious disease control efforts operating without political awareness and expediency may be idealistic, but it is also naïve—sometimes, to the point of inviting disaster. Epidemic control efforts have to adapt by explaining to the politicos that it is a health-education, health–promotion program, or human-rights gathering—not fomenting revolt. Global health thus has to learn to change its messaging to face the inexorable; to bow to the invasiveness of bigger pictures, broader realities (Fig. 2.6).

Adaptability to politics, economics, and to the wealth and poverty of the world: health programs, in emergencies or over the longer term, need to be aware of local incomes and costs of services—and how they can be paid for [5]. Programs and projects and interventions should thus evolve to make sure the treatment, the testing, the assistance they are promoting is available and affordable—both when the internationals are there, and after they leave. Programs should also perhaps adapt economically—to provide access, and to sustain. Economic adaptations are necessary, as well, to integrate health with opportunity—with jobs and dignity. Infection support groups—for anyone who has tested, whether their result is positive or negative—can be linked with market gardens or micro businesses: buying hens, selling eggs. Such lo-fi economic schemes can, in turn, link back to encouraging locals to attend counselling and testing, or becoming a member of a support group. All of this creates benign loops.

Figure 2.6 When infection control caravans break down, put up tents instead: adapting program branding, logos and language to local needs in Tanzania. *Picture courtesy Sebastian Kevany.*

Back in Tanzania—in Kisarawe, outside Dar es Salaam—I was involved in income-generating programs that brought outcasts back in, that integrated the ostracized. This was integration for health benefits and for morale; integrating to sidestep the stigma of infectious disease; integrating to address the consequences of locals who have been banished from jobs or homes because they were infected, or suspected to be infectious, and who had no other way of surviving. The programs facilitated locals building chicken coops adjacent to counselling and testing tents: for the internationals, it meant the pleasure of walking through verdant gardens tended to by HIV/AIDS patients. It meant seeing communities, down long, winding jungle roads, regrouping in support of the damned: meant seeing them regain a livelihood, nutrition, and a sense of worth as they leaned on their shovels and greeted you. This was truly enlightened, renaissance global health—adapting to build morale amongst the sick, the marginalized.; curing economic woes, as well as medical ones (Fig. 2.7).

Figure 2.7 In South Sudan, near the Ugandan border, dinner could be hard to come by — requiring adaptation to diet, as well as everything else. *Picture courtesy Sebastian Kevany.*

2.5 Key messages

- Adaptability in the context of global health diplomacy is relevant to the individual as much as to the design and delivery of the program
- In terms of epidemic control, tailoring approaches to local cultures, norms and beliefs can make all the difference between success and failure.
- Adaptability has to remain within the realms of the possible and also within the boundaries of formal intervention protocols.
- Adaptability is dependent on local involvement and input, via community working groups or other mechanisms, that give recipients a voice.
- Though adaptable programs that are nimble enough to respond to local conditions, international relations and security are augmented (Fig. 2.8).

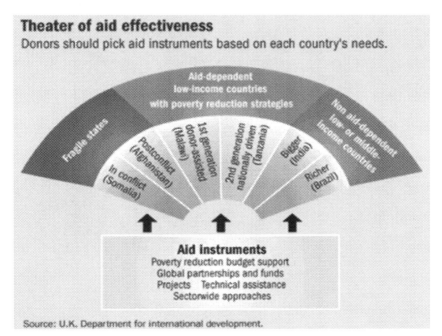

Figure 2.8 With so many different realms of global health efforts, adapting to different paradigms while also maintaining fidelity to core goals and protocols is crucial.

References

[1] Kickbusch I. Global health diplomacy: How foreign policy can influence health. Br Med J 2012;2011(342):d3154. Retrieved from http://www.bmj.com/content/342/bmj.d3154.

[2] Kevany S, Sahak O, Workneh NG, Saeedzai SA. Global health diplomacy investments in Afghanistan: Adaptations and outcomes of global fund malaria programs. Medicine, Confl Survival 2014;30(1):37–55. Retrieved from http://www.tandfonline.com/doi/pdf/10.1080/13623699.2014.874187.

[3] Kevany S, Khumalo-Sakutukwa G, Murima O, Chingono A, Modiba P, Gray G, et al. Health diplomacy and adapting global health interventions to local needs. BMC Public Health 2012;12:459. Retrieved from http://www.biomedcentral.com/1471–2458/12/459/abstract.

[4] https://www.dictionary.com/browse/adapt.

[5] Kevany S, Murima O, Singh B, Hlubinka B, Kulich M, Morin S, et al. Socio-economic status and health care utilization in rural Zimbabwe: Findings from Project Accept (HPTN 043). J Public Health Afr 2012;3(e13):46–51.

[6] Katz R, Kornblet S, Arnold G, Lief E, Fischer J. Defining health diplomacy: Changing demands in the era of globalization. Milbank Q 2011;89:503–23. Retrieved from http://onlinelibrary.wiley.com/doi/10.1111/j.1468-0009.2011.00637.x/abstract.

[7] Irish Department of Foreign Affairs. (2014). Review of Ireland's foreign policy and external relations. Department of Foreign Affairs White Paper. Retrieved from https://www.dfa.ie/aboutus/what-we-do/our-strategy-and-guiding-principles/foreign-policy-review/.

[8] Kleinman A. Four social theories for global health. Lancet 2010;375:1518—19. Available from: https://doi.org/10.1016/S0140-6736(10)60646-0.
[9] Kevany S, Benatar S, Fleischer T. Improving resource allocations for health and HIV in South Africa: Bioethical, cost-effectiveness and health diplomacy considerations. Glob Public Health 2013;8:570—87. Retrieved from http://www.ncbi.nlm.nih.gov/pubmed/23651436.

CHAPTER THREE

Power to the people: local ownership of infectious disease control

Abstract

In the post-colonial era, the concept of country ownership and local ownership of infectious disease or other global health programs has increased in prominence and importance. While in many settings this is related to the transfer of funding and resourcing to local actors in order to reduce dependence on aid, from the barefoot diplomacy perspective it is also critical that local ownership engenders a sense of community investment—of pride in the program. This, in turn, relates to questions of sustainability and transferability of health responsibilities to local actors—all of which need to be managed with the principles of diplomacy and international relations in mind. The techniques by which this can be achieved—ranging from the development of local community advisory boards, to the certification of local ownership of programs, to the solicitation of other sources of local advice and expertise on what program will function best in each setting—are explored here.

The MLI model to advance country ownership

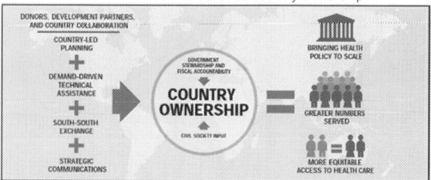

Figure 3.1 One of the many different aid and development models that advances the concept of local ownership.

3.1 Local ownership

It's easy, in life, to forget there are many varying points of view. In global health, it's easy to forget the views of locals, of target populations, of aid recipients (Fig. 3.1). It can be hard to imagine what it all looks like through their eyes: is the program good or bad; feared or welcomed? Can it be made to work for the locals, on different levels, as much as for the internationals? If not, is designing and delivering infectious disease control efforts a game that only the internationals, the global West and North, can play? These days, as it always has been, the world is governed by the haves; having such views is an automatic reflex, an assumption, a *fait accompli*. Yet it's so easy to forget that in the 20th century so much may have gone the way it did because of North and South, us and them, first- and third-world mindsets. Those mindsets are often wrong-headed, in terms of health and development as much as in terms of power, control, and program design.

Such are the mistakes of history; but today, in the 21st century, the world operates on a more equal, global basis. We are, we hope, more equitable and inclusive; less didactic and segregated. There is more collaboration, information sharing, exposure, and knowledge: in turn, locals in places with epidemic or health system problems are engaging more and more in donor structures. Locals are now, for the first time, engaging in the mechanics of "infectious disease control on an equal footing with the internationals—in ways that sidestep exploitation, giving the locals the same skills of diplomacy and finesse as the internationals." They are both thus engaging in a way that ensures locals can also pursue simultaneous, smart agendas and priorities through global health: engagement that helps poor countries and poor people to benefit from changing landscapes of style, power, and control. Engagement that, also, benefits by leveraging local style, local needs, and therefore building local control and ownership.

Local control and ownership means, most simply, that locals have a sense their responsibility for, and their connections with, global health programs. These efforts become locally owned—in terms of design, management, and style. External imposition is minimized, and instead there is country ownership: this involves empowerment of local actors to make decisions through diplomatic discourse, creating systems of local responsiveness and responsibility [1]. Country ownership of systems and interventions might thus be seen as dubious: you are asking locals to pay more, because they own it. However, it is also a sublime idea; it's the only way to make global health programs click—to make them last, succeed, and flourish. It is an epidemic control or a public health emergency response working because it is built by locals, for locals.

Local ownership is also a way to hear new voices. Poor countries, poor places, poor people with intelligence and insight, don't always get places at the intervention design table. Traditionally, they have had no place and no voice; but gradually, with the right energy and the right trust, locals are claiming their place, finding their voice. Finding the confidence to take responsibility and leadership; they are fitting desired results—fewer HIV/AIDS infections or more screening for tuberculosis—with local knowledge, preferences, and needs. They are also negating overbearing international styles: they are standing up to didactic styles that say, this is cost effective; this is our solution, our choice for your environment, society, and culture. By giving the locals a voice from day one, local control of programs can therefore help dodge the pitfalls of international dogma, inappropriateness, or imposition.

Locals also need an equal voice in program design through discussion, engagement, and idea sharing—instead of having distant or transient internationals calling the shots (Fig. 3.2). International staff need to learn from the

Figure 3.2 In Thailand, epidemic control education sessions were often combined with local village cultural displays. *Picture courtesy Sebastian Kevany.*

locals what the environment is like—details on local politics and security; what the local culture is; and other nonhealth needs. International staff also need to learn what else locals are trying to resolve or improve, in places where they are implementing medical interventions. Is Boko Haram, or the Lord's Resistance Army, also there—hiding out in the jungle? Is global health supporting them by mistake, or is it helping to win that fight as well via challenging isolated extremism through national and international presence and education? Is an infectious disease program changing perceptions, or providing places of refuge? Locals can thus design epidemic response programs that also recognize broader local realities: bigger local pictures, that kill two birds with one stone.

Local voices can also help internationals to adapt and to design culturally, socially, religiously appropriate programs for the region, the district, or the village: locals know what the villagers are going to accept, and they know the resources that are available. They know, also, how global health programs can best function in places where people are everywhere—but equipment, including technology, electronics, maintenance, or power is nowhere. The locals know how to respond to ideas from grass roots; they understand the role of devolution, community empowerment, village working groups, and advisory boards.

Such voices in the international development realm can minimize bad ripple effects and accentuate good ones; they can ensure that locals also get downstream effects from a program, beyond health. In Zimbabwe, I saw local empowerment working in favor of both locals and internationals; building feedback loops and communications, while improving efficiency and effectiveness. Saw internationals tuning in to local conditions, a world away from unthinking aid disbursements and distant technocrats. It was a step towards building allegiances between locals and internationals—both individual and national—that only trust and responsibility could inspire.

It was a step, also, towards *beau ideals* of stewardship, accountability, transparency and demand-driven global health: a step towards less corruption, and towards locals having a louder, clearer, more resonant voice in the design of the programs they received—programs they are meant to benefit from. A step towards collaboration, equality, loyalty, and better health. Our next essayist, Erastus Maina, describes the critical importance of such attitudes in a Cambodia still recovering from it's communist era.

Cambodia: A health system emerging from devastation of the Khmer Rouge regime

My current role is Program Manager for Safe Surgery 2020, which seeks to improve access and quality of safe surgical and anesthesia care in low- and middle-income

(Continued)

(Continued)

countries. We do so by: (1) advocating for increased prioritization of surgery and supporting health ministries in national surgical planning processes; (2) developing and mentoring surgical workforce including surgeons, anesthetists, OB/GYNs, and nurses; (3) investing in innovation development to improve surgical care; and (4) conducting rigorous monitoring and evaluation to understand and disseminate insights and advocate for further improvements in safe surgery. Safe Surgery 2020 is funded by GE Foundation, hosted by Dalberg, and implemented by four lead partners.

According to the Lancet Commission on Global Surgery report (2015), South Asia is one of the regions with significant unmet need for surgical care; 97% of the population in South Asia do not have access, compared with 36% in higher-income regions. To determine viability and scope of Safe Surgery 2020 programs in South Asia, we undertook the first of a series of scoping visits to South East Asia in April 2018. The objective was to assess demand for the program, identify countries that demonstrated interest to partner with us, and seek out potential partners that could complement our efforts. Our team visited Singapore, Laos, and Cambodia; I was involved in the visit to Cambodia.

Before the many meetings we held with different stakeholders in the Cambodian health system, our host briefed us on Cambodia's recent history, including the devastation of the country during the Khmer Rouge regime between 1975 and 1979. He especially emphasized the importance of acknowledging this conflict period in our conversations with government officials and other stakeholders, to better engage with them. This turned out to be accurate, because this issue was brought up by the country stakeholders in all meetings we took, as a way of explaining the state of their health system. It was reminiscent of my experiences in Rwanda, where the 1994 genocide is often readily acknowledged in most conversations.

The devastation of the physical infrastructure was evident within the capital, Phnom Penh, and even more so outside the capital. While some few of the hospitals we visited were shining models of modernity with advanced diagnostic equipment, ample numbers of trained physicians and health workers, and offering diverse specialties of care, a lot of other facilities were dilapidated, ill equipped and inadequately resourced. For instance, just a 2-hour drive outside Phnom Penh we visited a provincial referral hospital that demonstrated stark infrastructure gaps: the facility had only one operating room (inadequate for a referral facility), which was both poorly equipped and struggling with basic processes such as ensuring sterility in the operating room (OR).

Another striking issue was the shortage of trained physicians to man the public health facilities. We were told that a significant section of the educated population fled the country during the conflict years—and did not return (Fig. 3.3).

(Continued)

(Continued)

Figure 3.3 A difficult working environment. *Courtesy Erastus Maina.*

Western educated Cambodians who remained, including doctors, were targeted during the genocide—thus leaving the country with very few doctors to provide services, and train other doctors. Most trainee doctors now prefer to train abroad in Thailand, Vietnam, and Singapore where they can get better quality of training. Inevitably, some never return to Cambodia after their training.

Also adding to the strain at the public health facilities, approximately 2/3 of the health practitioners participate in dual private/public practice. In the rural areas, for instance, it is estimated that only 15% of medical practitioners are working exclusively in the public sector. In one of the facilities we visited (after 3 pm), we were not able to meet any of the doctors because they had left to work in their private practices. It is not hard to imagine the potential for conflict of interest; the dual practice could be a disincentive for practitioners to improve the quality of care at public facilities, since they would compete for the same patients at their private practices.

Today, the ministry of health, individual institutions, and hospitals have established relations with more advanced neighboring countries such as Thailand and Singapore, to support efforts to improve the health system. This is being done through direct funding support, learning and capacity building, and mentorship. The long-term ambition is to model the Cambodia health system on the Thai and Singaporean health systems. Yet my visit to Cambodia painted for me a picture of a low-income country that is still trying to emerge from the ravages of a conflict that all but decimated the health system.

Erastus Maina

3.2 Locally led success

In Zimbabwe, it became clear that in oppressive, politicized environments locals—with so little money, or contact with the outside world—often just wanted a distraction. They wanted entertainment, parties, interaction—as part of HIV/AIDS programs. They wanted a cool and inclusive scene that, in a way, they owned. And they owned it because they had asked for it; because, along with better health, that was also what they needed. The locals thus saw that entertainment also challenged stigma, broke down barriers, and built confidence to go to clinics, tents, or caravans and test themselves.

Global health programs that combined an entertainment element ultimately built rapport, trust, and connections between locals and internationals: these connections answered questions about how to justify epidemic control investment, and questions about the returns: these connections allowed for the right programs, designed by locals and internationals together, to operate optimally on multiple levels—and built the case for support.

That case is challenged, of course, by skeptics; by public and private, social and political doubts around the affordability, virtue, and value of global health investment. It is challenged by the skepticism of those who say that epidemic control investments only ever produces a dependency culture—meaning dependence of the locals on the internationals. This skepticism says that epidemic control makes life worse for poor people; that it stifles creativity and crowds out native efforts. It says that assistance makes locals lazy, indolent, and careless; says that it builds a welfare world, a form of neo-enslavement. This is a skepticism, that, in turn, has to be challenged as much as it challenges; that has to be defended against, as much as it attacks. Defended against, perhaps, by programs that locals own—that sidestep perceived dependence of the third world on the first world. Programs that dislodge the dominance—submission nexus; programs that promote, instead, independence and interdependence; that build and advance local prestige and dignity, and that challenge skepticism.

Internationals who tune in to local control can thus help in building increased capacity, confidence, empowerment, and global presence of in-country actors—not to mention improve public health emergency or epidemic response efforts [2]. Tuning in to local ownership is therefore the antithesis of dependency culture—it is a refutation of dated arguments against global health; a riposte to narrow utilitarian measures of effectiveness. Locals challenge dependency by moving themselves away from the image of lacking confidence, capacity, or dignity; local ownership gives locals active, confident, and empowered roles by moving away from critical perceptions of locals as passive recipients of money—away from images of locals following the Pied Piper.

Local empowerment also challenges didactic international styles. It challenges the paradigm which says, "This is what you need, so this is what you get." Such styles, today, are so undiplomatic and can be in such bad taste—not to mention ineffective—that in and of themselves they can undermine epidemic control responses, making them neo-colonial and prescriptive (Fig. 3.4). Local knowledge trumps international didacticism, producing better outcomes for both health—and for respect, egalitarianism, and equality as well. This is the diplomatic, helping to ensure that the countries, places, and locals receiving health assistance gain more control—and are considered equals. It is the diplomatic as the equitable and the progressive; the enlightened as an antidote to skepticism—skepticism also, from the locals, as much as from the internationals. Local skepticism is thus a reaction against global health veiling neo-colonialism, while local ownership ensures that locals will not be colonized again.

Figure 3.4 Also in Thailand, local design of testing and education centers was critical to program utilization and acceptance. *Picture courtesy Sebastian Kevany.*

So local ownership, like diplomacy itself, is abstract, theoretical—but it is also real. A Nigerian student once told me about internationals digging bore holes and building hand pumps and wells for clean water access in remote, dry, villages: these clean-water bore holes and pumps were never used. The locals, not trusting them, didn't use them—didn't maintain them, when they broke. They didn't care if pump handles were stolen; not having built them, they didn't feel a sense of ownership. Ownership only came much later, with input and involvement by tribal chiefs, elders, and politicians—and also by washer women, housewives, children, and the man in the street.

Troubleshooting the issue was coordinated through groups and forums—through creating systems for local, legal, and symbolic ownership of the bore holes, with certificates, titles, and codes of practice. This resulted in locals feeling proud, feeling connected—feeling ownership—and the careful maintenance of their new asset. With this, as Julia Chang and Stephen Straube illustrates in our next essay, cohesion and coordination will enable even the most challenging global health efforts

Emergency care—unexpected basics in teaching the ABCs

Emergency care is crucial to limiting the morbidity and mortality of injuries and other acute medical conditions. Global health efforts, in decades past, have been largely focused on treating and preventing infectious disease, but more recently, the need for access to emergency care has become clear. For patients with life-threatening, acute presentations of traumatic injury and infectious disease, as well as for many other increasingly common diseases, such as heart disease, lung disease, and stroke, and of course in public health emergencies, emergency care can be the difference between life and death, survival with good outcome and long-term disability.

A first step to improving access to emergency care is identifying the specific needs of different settings. In 2018, the World Health Organization (WHO) launched the Global Emergency and Trauma Care Initiative, to help identify gaps in national emergency care systems and implement interventions to address these gaps.

One notable gap demonstrated in prior research is the lack of health-care providers trained in emergency care. To help address this, we recently developed and piloted the use of a mobile-phone accessible educational tool based on the WHO Basic Emergency Care course. The tool comprises a suite of clinical cases capturing common emergency medical and trauma conditions.

The development of this tool was guided by a few key criteria. First, we wanted to harness the power and widespread use of mobile devices. Furthermore, our prior experiences led us to look for platforms that would make this tool freely accessible online, useful across various operating systems, and requiring as little internet

(Continued)

(Continued)

bandwidth as possible—for easy downloading, in locations with slow access to data connection.

A team of colleagues collaborated to write a suite of 32 clinical case scenarios. Scenarios were chosen to represent a broad spectrum of emergencies, such as road traffic injury, lung infection, and pediatric snake bite. High-yield learning points were incorporated as multiple-choice questions, walking learners through basic emergency medical decisions. Early in development, these clinical cases were piloted at a global health conference, and learner feedback incorporated into the final suite of cases.

This educational tool was formally piloted in Tanzania in the spring of 2018. Since then, it has been used in conjunction with the rollout of the WHO Basic Emergency Care course in a number of countries.

A few lessons that we have learned from the above experiences are worth sharing. First, during development of the tool, we quickly realized that small technical challenges can become major hurdles. Finding a format for cases that was engaging and interactive, without requiring significant internet data connection, was an early hurdle. After trialing many formats, we opted to use a storytelling format that allowed for incorporation of images in addition to text, as well as empowering the learner to make active choices, while keeping file sizes as small as possible.

Another early lesson came from challenges in finding a host for the website that was built to offer access to this clinical suite of cases. We used a well known, reputable public company. Unexpectedly, one year later, during the deployment of cases to a large number of learners, we learned that the website hosting company had decided to block access in many developing countries due to concerns about fraudulent activity. Another website host was required in short order.

Finally, many new needs have been uncovered by discussing clinical cases with frontline medical providers as they use the educational tool to learn emergency care. For example, one scenario recommends providers in clinic settings offer patients who are very short of breath oxygen by nasal cannula and transport to higher level of care. A number of frontline providers pointed out that oxygen and nasal cannula were not available in their clinic settings. Their feedback led to very productive discussions during the Basic Emergency Care course.

In sum, the wide use of mobile phones and availability of data connection has opened new doors in local-level medical education. Paired with the urgent need for emergency care training in many parts of the world, we were part of an effort to develop a novel educational tool for frontline providers in developing countries to learn basic emergency care. Lessons learned included the importance of being flexible and creative with available technologies, as well as actively incorporating learner feedback.

Julia Chang and Stephen Straube

3.3 Local protocols

These examples illustrate that active ownership requires different skills than passive acceptance does. It requires diplomacy and finesse—it requires the smart and the cool to make locally owned global health function optimally. It has to realize local dreams of getting the right service, in the right place, at the right time—as well as international dreams, of reaching targets and development goals. These aspirations implicitly involve remote villages being given a voice in epidemic responses—distant villages, in the jungle, being able to negotiate and to express—to have a means of expressing ideas that distant directors may never have thought of. To achieve this, the locals need capacity: there is just as much a need for streetwise, multilevel, enlightened strategizing in southern villages as in northern capitals.

Enlightenment on the issue of program ownership will also, for the local as much as the international, help ensure that programs don't inadvertently generate unforeseen consequences: training and awareness can ensure that locals and internationals see all of the downstream effects of epidemic responses—the potential, as well as the actual. Enlightenment thus means locals operating on the same level as the internationals in designing, planning, and implementing projects; it means listening when locals say a certain part of a tuberculosis screening program, for example, isn't going to work. And it means listening when locals say they are going to like this malaria prevention program over the other, because it also improves environment, trade, and the economy, as well as health. Enlightenment ultimately means confident locals, no longer just waiting or watching from the sidelines, who are increasingly setting the tone, advising, calibrating, managing responses, and helping to govern.

Locals can in many cases become both *de jure* and *de facto* global health governors. In many places, I have seen locals leading barefoot diplomacy approaches—making sure that programs were leveraged for every possible local and international benefit. These people ensured that epidemic control was framed in terms of the direct—but also of the downstream; in terms of both narrow targets and broader realities—the economic and the political, the strategic and the diplomatic. Local control and international assistance thus operated together, resulted in better health outcomes, and indeed in better outcomes in areas beyond health.

Throughout the history of global health, there have been inevitable links between programs and resource extraction, often wrapped up as well with cold war politics or hidden agendas. Infectious disease control has often involved subtle and implicit strategic, economic, and security agendas—not just pure altruism. This hasn't always involved dark cynicism, either. Is the possibility of internationals providing aid just to win allies—or to prevent global epidemics from reaching their shores, or to stabilize spheres of global influence—something to be ashamed of; or is it, instead, what makes the world go round? Is such *realpolitik* still better than hard power—des the sick child in the mud hut at the end of the dirt road in the smoky village care if you are ultimately there out of pure altruism?

Local ownership operates in the same *quid pro quo* world—but without the intrigue (Fig. 3.5). Ownership, from a local perspective, merely makes the implicit explicit. Locals and internationals thus jointly pursue bigger pictures through the medical; local control allows both sides to pursue other pressing agendas through health—issues including security, diplomacy and economic growth. More local control means less need for hidden agendas; barefoot diplomacy then becomes more explicit, open, and inclusive.

Figure 3.5 In rural areas, such as this village in Tanzania, locals are often pleased to receive infections control aid, independent of the broader political or strategic context. *Picture courtesy Sebastian Kevany.*

Why open? Well, if global health has to be an exchange, a *quid pro quo*— if that element is inexorable, in the name of achieving bipartisan support for such efforts—then it is better perhaps, for both locals and internationals to accept that—and to optimize the consequences. It is better, in other words, to optimize the downstream benefits of local power—which outweighs the risks of local control and ownership. These risks are consequential on local ownership of global infectious disease control, and on international health security being devolved to the locals: these are grave local responsibilities.

Local control, ultimately, allows global health to work better. The mitigation of the risks it brings depends on local skill and capacity in epidemiology, program design, and diplomacy as much as funding, commitment, and political will. Local control is thus dependent on a transfer of responsibilities that were once the exclusive purview of the internationals: In recent years, international HIV/AIDS, malaria, tuberculosis, and other epidemic response efforts have been deemed an international success—as a triumph of diplomacy, of altruism. While such efforts can indeed be regarded as triumphs, without local inputs they are also intrinsically failures.

As the saying goes, "no good deed goes unpunished," whereby [3] some worry about sustainability—about expanded treatment of epidemic care coming at a cost to patients who bear the burden of other chronic diseases [4]. But both local and international skeptics are mainly concerned about funding and support—about "crowding out"; about prescriptive inappropriateness. They are concerned about short-sightedness that for every new patient on antiretroviral drugs, further unexpected support will be required. They are concerned about the side effects of medication support—about cost of adherence to the daily regime [5]. They worry also about ensuring patients keep to drug plans to avoid drug-resistant illnesses developing—and spreading. The concern shifts, in such situations, to preventing the emergence of new viral strains, wily, and resistant enough to outwit the current drugs.

These are the ripple effects of infectious disease control programs that often slip under the international radar, but are all too clear to the locals. The locals already see these traps and pitfalls of sustainability and coherence that the internationals would miss. While working with in a hospital in the Manenberg Township in Cape Town, South Africa, the same dominos started to fall in my mind. There, ironically, internationals were being cast as villains, with locals saying "you started us on antiretrovirals." If patients can't continue, or aren't treated for side effects after the inital surge response— there will be trouble, on diplomatic or international relations levels.

Bigger trouble, perhaps, than if you had never come in the first place: "No good deed goes unpunished." Even with the best intentions, without the involvement of locals—who see problems coming over the hill and around the corner—the potential for international incidents as much as for discontent, resentment, and disappointment exponentially increases. It is only when locals and international treat each other as equals, in every context, that global health programs can work—as our next essayist, Dr. Karen Weidert, demonstrates.

Call me Jobu: A tale of pregnancy and miscarriage in Ethiopia

I started my worldly travels at age 20, when I had little inhibition and a lot of adventurous spirit. For me, traveling made the world feel smaller—I could always find ways to connect with people, despite having no shared language or culture. Years later, global public health seemed a natural fit for my interests. I had long forgone aspirations of medical school, after deciding I wanted less school and more experiences. I thus completed my master's in public health at University of California, Berkeley and agreed to remain at the university for one more year in a staff position—at a research center which focused on improving access to sexual and reproductive health in sub-Saharan Africa. I agreed to only one year, as I felt strongly that academia was not for me, as it felt too disconnected from the work on the ground. One year quickly turned into two, and then three. But, I was doing what I wanted, so it didn't feel like a trade-off: I was managing a three-year research study to test a new model for increasing access to contraception in northern Ethiopia, traveling to the field several times a year.

I had been in east Africa for over three weeks, recently arriving in Ethiopia from Tanzania, when I started to recognize unusual bodily responses to what was a typical monitoring trip. I was used to spending 12+ hours in a 4 × 4 crossing rugged terrain, and trekking through the heat, to find community health workers. It normally did not bother me. Today, waves of light-headedness would wash over me every time I got out of the truck and my stomach felt off—but I reminded myself that stomach issues are pretty standard when I am traveling in rural Africa.

Eventually, I confided in an Ethiopian colleague that though I was 99.9% sure I was not pregnant, I was frustrated with the general nauseous and lethargy I was experiencing. I was embarrassed to discuss this, given my "expertise" in reproductive health. She quickly eased my shame, but stated she was not 99.9% sure, like me—and a few days later informed me that she had made an appointment for me at the rural health center we were monitoring. I thought this was funny, since my experiences while conducting monitoring visits to such centers was that you did

(Continued)

(Continued)

not make appointments—you just showed up, waited in line, and got the services you needed, if they have them that day. Nonetheless, I obliged to the test.

Upon arrival at the health center, I needed to get a registration card. It wasn't busy, and several staff were very excited to help with my paperwork. With the help of my colleague translating, I was asked my first name (Karen), which when written looked to me like "Heren". Instead of stating my last name, I just automatically started spelling... "W". This was immediately written down as "Jobu" and the name portion was deemed complete: Heren Jobu would forever be the record of me in Ethiopian health care system.

The rest of the process was methodical, and it didn't feel all that different from my experiences with health care at home—in fact, the time between testing to results was much more efficient. After I was registered, I was moved to another room where a form was completed to designate the services I needed. Then, I was escorted to the laboratory. Through a window with metal bars, I was handed a little cup with no lid and told to fill it up halfway in the pit latrine. I walked over to the pit latrine to be greeted by family of wasp-like insects swarming as I squatted, and a smell was that was exacerbating the baseline nausea I had become used to over the last couple of weeks. I set my cup in front of the window; the technician slid rusted tongs through the bars to retrieve it and pointed for me to sit down on the bench with the line of other patients waiting for results.

Within 5 minutes, he called my name and I returned to the window. He very proudly explained to me that I would not have to pay for the test because services were free, even for a *faranje* (foreigner), but I needed to pay seven birr for my registration card. This was definitely an improvement from the U.S.: I said, "Great—but what is the result?" He replied, "Positive, of course." Yes, of course.

I realized in that moment how all of the signs had been there all along, and I was just in denial. As a researcher with a focus on family planning and contraception, I recognized the irony that I was dealing with my own unplanned pregnancy. I now had first-hand understanding of unintended pregnancies, contraceptive method failure, and ignoring signs of pregnancy: these were all things I had written about at length in proposals, manuscripts and reports. But in my writing, those were things that happened to other women—like the Ethiopian woman behind me in line at the health center, who likely faced many barriers to accessing effective contraception, and might have received the same test result right after me.

(Continued)

(Continued)

Figure 3.6 Karen, far from home—but made to feel at home. *Courtesy of Karen Weidert.*

Except it was me, and she and I were really no different (Fig. 3.6). However, my consolation was that though my pregnancy was not planned, it was wanted, which is not the case for so many women, maybe even the woman behind me in line. And then what would she do? I didn't know the answer, but the fact that she may not have all the options I did reaffirmed my commitment to my work in reproductive health.

The story could have ended here. I was due to go back to California in a couple of weeks, but I was heading back to the town center in a couple of days. There, I could get an ultrasound from the OB-GYN who was the coinvestigator on our study and had a private clinic. When I got back from the rural monitoring, I set up a meeting with him to provide feedback on our findings. At the end of the meeting, I mentioned the positive pregnancy test to him. Again, I was embarrassed that I was in this situation and also that I had to discuss my personal life in our otherwise professional relationship. He brushed it off, and requested I come by his clinic later that evening for an ultrasound, which I did. As he conducted the ultrasound, I could tell that something was wrong. He asked me when my last menstrual period was, and noted he would have expected more development.

(Continued)

(Continued)

A week later, he had me come back for a repeat ultrasound, which confirmed the worst. As I sat in his office, he gave me the news that I would likely miscarry, and provided me with information and ultrasound images to share with my doctor at home. I cried openly, as Americans do, and he consoled quietly, as Ethiopians do. In that moment, we were doctor and patient, but the next day we would go back to coinvestigators. Similar to a decade ago, when I embarked on my first international trip, in that moment, I was reminded the world is not so big. I was thousands of miles away from those I loved and navigating a difficult experience —but I wasn't alone. I just had to be willing to look beyond language and culture, to shared experiences and common humanity.

Karen Weidert

3.4 A confident local voice

Without accessing local *nous*, epidemic control or other health efforts are skating on thin ice. Local knowledge is dependent on local confidence; harnessing this can generate tailored health responses fine-tuned to local needs. Both internationals and locals thus need to understand that global health can't operate without fair, equal partnerships—partnerships that allow the locals in the door; partnerships that leverage local expertise. Smart approaches need to employ the locals who know what to do—needs to harness local intelligence, so that international experts no longer operate in isolation. Internationals and locals can then, together, use what is available on the ground—local resources, inputs, people—to produce programs better attuned to cultural, religious, and social sensitivities—and, ultimately, result in better health. The result is, above all, better health and intervention results because locals know which projects other locals are going to like, trust, and embrace.

Locally designed global health will result in building programs that are sustainable, while also building and supporting other systems that can advance multiple agendas. Put simply, building programs that locals trust, are more likely to succeed—the ones that develop local confidence, capacity, strength, and empowerment.

Anyone can see that—on any day of the week, out in the field—the fundamental skill—the inherent intelligence, knowledge, understanding—is there amongst the locals. This knowledge is part of the fabric of society—part of the environment, experience, culture, DNA, and blood memory [6]. But local intelligence emerges only when it is allowed to, for example as in the tales from Zimbabwe or Tanzania: in remote, breeze-block community centers, under jacaranda trees. In community working groups—in forums for the reticent, the shy, to speak and to express their views: the locals' insightful, critical views on innovating, adapting, and advising. Views that see around corners, that anticipate, as the locals say, "challenges" or "speed bumps." The knowledge that describes challenges that internationals wouldn't and couldn't have foreseen: the locals know every detail of hidden traps that would otherwise derail the best of intentions. They can provide insight and awareness that those in the distant centers of the global health governance universe could only dream of.

There are, then, so many different benefits to local control of health programs. Local understanding of the direct and indirect potentialities, of the collateral and the downstream effects—of the risks and benefits of epidemic responses—leads to an un-complicated, simpler process. Local control starts both the process of simplifying partnerships between internationals and locals, and uncomplicating suspicions, resentments, or post-colonial angst. Such involvement also improves transparency; it allows for a showing of cards, reducing the need for negotiation and wheeling and dealing. It facilitates an end to exploitation, trickery, and the fear of being taken for a ride; it facilitates the beginning of downstream benefits to internationals, which is balanced by unexpectedly benign effects for locals as well.

Internationals also tune in to local control of global health as a gesture of trust; a symbol of willingness to devolve responsibility, triumph, and the spotlight. That trust dovetails into local pride—into the self-belief that—given the resources—poor countries, poor people, and poor places can address their own problems. Trust is essential. The absence of trust seems to be is the source of so many of our modern woes; its presence generates shared moral responsibility for the world and its future.

But the right conditions for trust are dependent on local capacity: the capacity to see the direct and indirect value of the right program, in the right place, at the right time (Fig. 3.7). Local capacity for processes, planning, deal making, and for aligning both the health and nonhealth agendas of locals and internationals. Smarter, cooler, renaissance global health is therefore often dependent on negotiation—on the locals and the internationals sitting at the same table. It is dependent also on locals operating on an equal basis as internationals: having the same skills, the same training. Equal training, in health diplomacy and other techniques; the same knowledge of the weaving

Figure 3.7 Slowing down Western pace to local speeds, and taking time to converse with staff, aids development of local voices and ideas. *Picture courtesy Sebastian Kevany.*

and the interplay of international relations, diplomacy, politics, and health. Equal knowledge of economics, security, conflict—and of the ripple effects their combination creates. But most of all the locals need to be at the same table as the internationals because if you are not at the table, as the saying goes, you are on the menu.

Local capacity gives locals the strategic, diplomatic skills they need: capacity and understanding are not necessarily only amongst doctors and politicians, but also amongst community leaders, villagers, and patients. To develop that capacity, that understanding, the locals need new skills, new forms of education. They require the confidence and belief necessary for them to take local control in barefoot diplomacy, epidemic control, or public health emergency efforts. All such efforts can demonstrate to local people that their efforts and investments in health also advance, in parallel, solutions to their nonhealth problems.

On my travels, saw the origins and the magic of local power far from the smoky, dusty villages; far from the jungle forest and dirt roads. But local power was also apparent in learning from experiences in my adopted home city of San Francisco, in the apogee [1]. San Francisco was avant garde, because it has had to be; because of its own HIV/AIDS outbreaks and epidemics that conventional responses couldn't solve. Its outbreaks and epidemics, within affluent borders, were divided by class, economic status, culture, and sexuality: local health fears in San Francisco at that time were a microcosm of current global health security fears: fears of epidemics amongst the have-nots, the peripherals, the fringe dwellers—which made the haves, the privileged, start to fear for their own health as well.

The haves ultimately, desperately, began asking the advice from the have-nots, connecting elitism and populism in true "gold rush" style. Connecting the affluent doctors, politicians and policymakers with the ordinary man or woman in the street—breaking down barriers, and empowering the fringe. The San Francisco elites thus began involving the disenfranchised by tuning in to community involvement and civil engagement: they began listening to everyone, from every background, in an effort to try to find solutions. Began listening, to try and understand what was going wrong, and how to set it right.

San Francisco was thus a prelude, a parallel—on an invisible scale; it was an example of local power. A prelude, as even if San Francisco isn't a country but a city; nor two countries—a rich one, and a poor one. Even if the haves in San Francisco can't be compared to internationals any more than the city's have-nots can be equated with poor African locals— the parallel is still there. There are parallels, also, in activism: in San Francisco's protests and marches bearing many similarities to later "treatment action" campaigns in South Africa. Parallels also in the process of local empowerment—in the need for a *soupçon* of the power of civil discontent to catalyze change. And parallels in the catalyst of the have-nots finding a voice; parallels in the once-marginalized citizens and communities becoming heard, confident, and empowered.

There were parallels, as well, in the smart effects that emerged from the activism in San Francisco—in the fact that local control of HIV/AIDS agendas and responses also resulted in other needs being addressed. Needs, not only in terms of health, but the bigger pictures of epidemic responses advancing equal rights—human rights. Advancing enfranchisement; advancing progressiveness, equality, and tolerance. San Francisco thus provided a template of smart moves, benign ripple effects via tapping into local knowledge and awareness—in terms of generating understanding (Fig. 3.8).

Figure 3.8 The essence of local ownership? Locals voice their suggestions for infectious disease control efforts in rural Zimbabwe. *Picture courtesy Sebastian Kevany.*

3.5 Key messages

- he thoughtfulness, insight and intelligence of local peel are often overlooked when designing and delivering epidemic interventions.
- This wakens the programs and projects on many levels, not least the utilization of services that may not fit with local customs.
- TGreater involvement of local people through community advisory boards and in other ways leads to the revelation of key local techniques.
- In turn, this results in scientific knowledge being augmented by local insights and wisdom, which helps in epidemic control.
- This also helps with sustainability and transferability of programs, as well as the transfer of knowledge (Fig. 3.9).

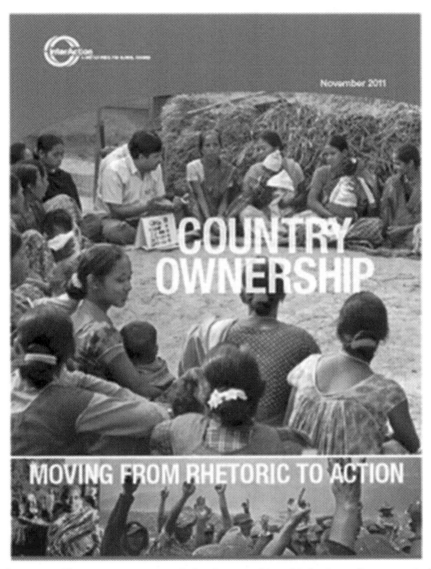

Figure 3.9 The movement towards local ownership of infectious disease control efforts is gaining ground worldwide.

References

[1] Kevany S. HIV/AIDS, *Global health diplomacy, and "San Francisco Values": from the local to the global.* Ann Glob Health 2015;81(5):611–17 Sep–Oct.
[2] Fauci A. The expanding global health agenda: a welcome development. Nat Med 2007;13:1169e71.

[3] Lyman PN, Wittels SB. No good deed goes unpunished: the unintended consequences of Washington's international HIV/AIDS programs. Foreign Aff 2010;89(4):74—5.

[4] Kevany S, Fleischer T, Benatar S. Improving resource allocation decisions for health and HIV programmes in South Africa: bioethical, cost-effectiveness and health diplomacy considerations. Glob Public Health 2013;8(5):570—87.

[5] Kevany S, Meintjes G, Rebe K, Maartens G, Cleary S. Clinical and financial burdens of secondary level care in a public sector antiretroviral roll-out setting (G.F. Jooste Hospital). South Afr Med J 2010;99(5):320—5.

[6] A phrase borrowed from Christian Beamish's Surf Odyssey. *The Voyage of the Cormorant* Patagonia Books (2013), in which the author refers to the resonance of instincts and intuition, across generations.

CHAPTER FOUR

A governance revolution: synergies versus turf battles

Abstract

Traditional demarcations between governmental purviews reflect those of academia—yet there is an increasing tendency, in both realms, to adopt those more holistic approaches which can produce synergies between initiatives and sectors. In the context of global health diplomacy, there are, at present, few formal ways in which considerations of foreign policy and international relations can interact with health, epidemic control, public health emergency responses, or other altruistic endeavors or policies on a structured basis. In this chapter, we explore ways that governmental polices related to all such spheres may be made to function in greater harmony through the development of appropriate fore and offices that link associated (previously disparate) outcomes.

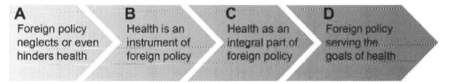

A	B	C	D
Foreign policy neglects or even hinders health	Health is an instrument of foreign policy	Health as an integral part of foreign policy	Foreign policy serving the goals of health

Figure 4.1 Epidemic control efforts are ascending the foreign policy spectrum.

4.1 Synergies and post-partisanism

There are, for me, so many words and ideas orbiting around global health: foreign affairs, foreign assistance, international development, aid, foreign policy, international relations (Fig. 4.1). These are vague terms, often appearing distinct, yet they overlap and are linked in nebulous ways. The reality of these links is increasingly evident—inescapably explicit. It is, today, impossible to separate the national and international functions of diplomacy such as world stability and security or international relations from epidemic control. Given the global nature of culture, economics, equity, defense, and conflict resolution, it is more and more difficult to separate them into different boxes, different compartments. Each bleeds in to the other in our interconnected and wired world, now more than ever.

Barefoot Global Health Diplomacy.
DOI: https://doi.org/10.1016/B978-0-12-818681-7.00002-9

In government, this "bleeding" occurs via a subtle, virtual revolution—through interdepartmentalism and interdigitation, through the interdisciplinary and the barefoot. This network reflects the way the world—academia, professions, individual lives, countries—is more connected now than ever in the omniscient, omnipotent 21st century. The interweaving of global health with bigger pictures of world security, environment, and diplomacy reflects the same changes and themes—the interdigitation of the global world. A world in which multitasking and multifariousness, instant communications and local and international awareness, are all part of our expectations: part of the *milieu* of the smart century.

This is a century governed by information; through maps of epidemics; through maps of weak health systems, staff shortages, and outbreaks [1]. However, health maps overlap with other realities as well—world trade maps, maps of the world's natural resources; maps of security and terrorism that show flashpoints and risks, maps where extremism anger and resentment originate. There is no single ruling map: all such maps are equally real. All are part of the same picture of coexistence, and all increasingly overlaid in the global networked world. Maps of disease, trade, resources, security, and aid can now be viewed together—and new connections and opportunities, synergies and solutions, suddenly appear.

The synergies of the pursuit of dual goals are part of the rationale for barefoot diplomacy. The rationale is built on links between global health and diplomacy; between the downstream effects of infectious disease control programs and policies on international relations—the links of security, peace keeping and stability along with the links to foreign policy and economics. Smart approaches, even in the epidemic control or public health emergency context, can play those parts and act in those roles. Without any such spin-offs, the case for global health is left only to fragile altruism: barefoot diplomacy approaches can leverage the increasingly inevitable interplay: can convince skeptics that altruism is worth more than mere charity; and far more than the generation of a global dependency culture, a world welfare state. Instead, epidemic control can be seen as essential—not only to health, but also to world security, functionality, and cooperation.

However, damaging, skeptical perspectives on global health responses are just as real and prevalent in the world as these idealistic perspectives. Such views are visible when narrow specialists emerge from bubbles with siloed, cubicle mindsets. As a counterpoise, barefoot diplomacy reflects a revolution in government and policy in the 21st century—wherein every department, every policy operates in synergy and is reconciled, balanced, and bridged to achieve consensus. Encouraging and facilitating such synergies is thus required to convince skeptics that infectious disease control is worth investing in: to benefit the international as much as the local, the right as much as the left.

Bipartisan support for global health therefore needs to be built on links between foreign affairs, global politics, and epidemics; the links between the interests of the global community and the nation state. These links, that grow stronger—more necessary, more immediate—each year are also links that allay the tensions between globalism and populism that the world is becoming increasingly aware of. How? Well, these links connect low politics—often associated with foreign assistance, aid, development—with high politics such as security, diplomacy, and economy. These links thus elevate concerns about the pursuit of a previously peripheral value (for example, the infectious diseases of the very poor) to prominent places on national and international stages (Fig. 4.2).

The integration of policies can also expand the reach and scope of infectious disease control efforts by catalyzing international involvement in previously peripheral places. This integration can be seen in countries and regions with public health emergencies or epidemics; can be seen in countries and regions that might not catch the international eye on that basis alone—that risk being forgotten about, without other reasons for international involvement. The ascendancy of global health can catalyze attention to, and assistance for, these forgotten places—all the way up the political spectrum—through recognition of diplomatic, economics, and international relations values and effects.

Global health's ascendancy in government, policy, and foreign affairs can therefore be described as a benign, 21st century phenomenon; the apogee of a 100 years ago, when international affairs were too often resolved by soldiers dying in the mud. The world has turned, over the course of a century, and has learned new lessons—new ways to resolve international tensions. This is reflected also in the declining public tolerance, in developed countries, for involvement in foreign wars. New ways of operating and maintaining world stability and security are also both demanded and facilitated by our increased links through communication, travel, and exposure; all of which make distant battles, epidemics, and tragedies both immediate and real.

New ways of dealing with such realities are thus demanded through a changing global awareness of the consequences of infectious diseases—through visibility of the manifest effects of and disease amongst the world's poorest people. pandemic efforts should perhaps therefore also be linked to health security concerns—to concerns of the threats and fears of epidemics

Figure 4.2 In presecession Southern Sudan, there was a significant overlap between nation building and epidemic control efforts. Here, UNICEF, the UNDP, and the global fund to fight AIDS, tuberculosis and malaria engage in joint governance. *Picture courtesy Sebastian Kevany.*

traveling across borders. These threats are, in turn, linked to looming, dark visions in the collective consciousness—of nightmarish outbreak scenarios permeating our daily fears, similar to a Hollywood thriller.

When working in Samoa and Vanuatu, I was lucky enough to see a subtle, implicit revolution in the way government, politics, diplomacy, and epidemic control could operate—both independently and together. In a small Pacific island nation, there naturally are lower walls, fewer barriers between these operations. There, I saw such distantly concocted ideas in motion: the health sector thinking of how it can advance other policy goals; also acknowledging that health policy can have a negative impact on other goals [2]. Thus there were smarter tuberculosis and HIV/AIDS programs that took opportunities to build a more equitable rural Samoa—an example of where responding to infectious diseases could have benign impacts on "bigger pictures".

So, barefoot health diplomacy, it is hopefully by now clear, can advance multiple benign agendas, which can then ease broader global tensions in ways that never sacrifice health gains or altruism (Fig. 4.3). Renaissance epidemic of health system strengthening responses can be consistently effective across departmental and ministerial lines—not by politicizing, but by "diplomatizing". This is a critical distinction, and one

Figure 4.3 In rural Samoan health centers, the links between broader rural development policies and public health efforts helped to limit rural depopulation. The sign reads; "No smoking, unless under the mango tree." *Picture courtesy Sebastian Kevany.*

that will reassure skeptics of a different kind: those who say that policy siloes and conceptual stove-piping are the way forward. Perhaps global health efforts that blur lines and reflects inexorable contemporary energies and trends are what is needed instead (Fig. 4.4).

Figure 4.4 A proud local nurse displays his wares in rural Vanuatu. *Picture courtesy Sebastian Kevany.*

In other words, epidemic or public health emergency responses reflecting international partnerships between North and South is arguably the pivotal dimension of bilateral relations: [3] advances in diplomacy are about the way you do things, as much as what you do. It is as about the way that equitable, accessible, tolerant, culturally sensitive styles, and approaches add up to a lot more than just new hospitals or refurbished clinics. Dr. Karolina Zielińska, our next essayist, demonstrates that, conversely, the medical can add up to more than just the altruistic.

Examples of interactions between health assistance and international relations: the case of the state of Israel

As a political science expert specialising in international relations in the Middle East, I am quite distant from the on-the-spot practicalities of aiding human health in developing countries. Still, while researching the particular case of Israeli development aid to sub-Saharan Africa countries as a part of Israeli foreign policy, I came across a couple of interesting examples of how health aid and diplomacy can crisscross and interact.

The Mulago National Referral Hospital in Kampala (founded in 1913) is a vital element of Uganda's public health system. Upon Uganda's 1962 independence, its reconstruction has been completed by Solel Boneh, a major Israeli construction company, (then) state-owned. This was a manifestation of a warm and fast-developing relationship between Israel and Uganda which lasted throughout 1960s.

The tides turned, however, in early 1970s when the new ruler of Uganda, Idi Amin, became on ally of Libya and aimed at eradicating Israeli presence from his country, leading to the break-up of relations in 1972. In 1976, an Air France plane hijacked by pro-Palestinian terrorists landed in Entebbe, Uganda: in the course of the hostage crisis, a British-Israeli national named Dora Bloch needed to be hospitalized. She was taken to ... the Mulago Hospital. After Israeli forces' raid on Entebbe, resulting in the rescue of the hostages, the Ugandan secret police—in an act of revenge—forced Bloch out from the Hospital; and shot and killed her.

Uganda and Israel re-established relations only in 1994. A new wave of cooperation, including in the field of development, followed. Among the projects, one example is a salient one—bearing in mind the above historical background. Interestingly, in early 2010s, Israel found itself again among the Mulago Hospital's partners. Israeli experts funded and supervised the construction process of the Mulago's new Emergency and Trauma Unit, making it state-of-the-art and prepared—also thanks to the Israeli-donated equipment and training—to deal with possible emergency situations and mass casualty events, as well as with day-to-day influx of critical cases. During a 2016 state visit by the Israeli Prime Minister Benjamin Netanyahu, joined by the Ugandan President Yoweri Kaguta Museveni,

(Continued)

(Continued)

the unit was dedicated to the memory of Yoni Netanyahu: Benjamin's brother, who commanded the 1976 Entebbe hostage rescue operation and was killed in the process.

The Mulago Hospital re-appears as a constant part of Israeli–Ugandan relations, yet there are also apparently more direct cases of Israeli–sub-Saharan health diplomacy. Since the 1960s, Israel—for geopolitical reasons—supported Ethiopia's battle against an irredentist province of Eritrea, supposedly precluding any chance of having positive relations with the restive region. Still, what happened in 1993, just days before the successful Eritrean independence referendum, was that Eritrean leader Isaias Afewerki suffered from a malaria-induced coma. He was then flown to the Hadassah Medical Center in Jerusalem, and treated there. Many believe that his personal experience was an important factor paving the way towards Eritrea's entering in to diplomatic relations with Israel.

Severe difficulties have apparently also been overcome thanks to health diplomacy in the case of Guinea. This Muslim-majority country broke relations with Israel following the 1967 Israel–Arab war. While certain informal contacts were maintained, the stalemate lasted until 2014. In that year, Israel extended comparatively large-scale aid to three West African countries suffering from the Ebola epidemic. This included aid for Guinea, which—aside from a share in UN Secretary General Ebola Emergency Fund, to which Israel donated USD 8.75 million—received Israeli help in the form of a dedicated, fully equipped medical clinic. This gesture has been described by diplomats as a key factor that led to the launch of negotiations on the re-establishment of diplomatic relations—successfully concluded in July 2016.

These three particular cases cannot, of course, be analyzed in isolation from the broader picture of bilateral and multilateral relations of the countries in question; they in no way define these relations. Yet, they seem to constitute interesting snapshots of the specific role which intervention in the field of health can play in relations between nations. While each such intervention constitutes an act of goodwill, it undoubtedly also works in conjunction with participant countries' interests. The cases of Eritrea and Guinea illustrate how health aid can play a role of a material and/or public diplomacy leverage, allowing leaders to follow through with their broader agendas. As the Mulago Hospital example demonstrates, a health aid project can constitute a real-life backbone for relations, despite disruptions resulting from politics-related turmoil. Health diplomacy also helps in building people-to-people bridges, where relations are otherwise dominated by geopolitical considerations.

In this context, an interesting question emerges about the possible aftermath of the assistance which Israel extended across the border to the population of the rebel-held southern Syria provinces. Medical aid was the most highlighted part of it; the informal practice of admission of wounded and ill Syrians into Israeli field

(Continued)

(Continued)

clinics and regular hospitals started in 2013. It turned into a systematic aid program by Israeli governmental entities, as well as NGOs. On the health side, it involved further patient admissions (around 4000 persons in total, including many children—and not including their assistant persons), construction and maintenance of two clinics inside Syria, and the supply of medicine.

The program lasted until mid-2018 take-over of southern Syria by regime-allied forces, and ended with evacuation of around 100 Syrian White Helmets—members of emergency rescue teams—and their families through Israeli-held territory to Jordan. This aid program seemed ground-breaking, since it involved a direct contact with—and even temporary admission of—a population of a country renowned for its decades-old, deeply embedded hostility. With a positive breakthrough in Israeli—Syrian relations highly unlikely at the time of writing, only time will reveal the programmes true impact and significance as a manifestation of health diplomacy.

Karolina Zielińska

4.2 Optimizing governance synergies

The benign dominos of the holistic in global health are thus just as powerful as the ripples of a narrow program focus—because what is the point in saving lives if there is more harm than good done on other levels? Holistic approaches, with both epidemic interventions and diplomatic and security and stability responsibilities across purviews and departments, are the apogee of utilitarianism and of functionalism. They are, in a way, Napoleonic—never making a decision, a choice, a move, without first considering every possible consequence, without trying to optimize every ripple. Holistic approaches therefore involve designing and delivering health programs, epidemic response, or public health emergency programs that are coherent across policies; that have only positive impacts on international relations such as world stability, environment, and security.

Policy coherence requires describing an alignment and harmonization between the arms of government and the legislature, ministries and departments, political lefts and rights; it is a *deus ex machina* for barefoot health diplomacy [4]. It involves also ensuring that the left wing knows what the right is doing; ensuring that global health also advances trade and environmentalism, and vice versa. It necessitates, above all, statesmen and women—politicians and decision makers—looking at every literal and metaphorical weapon in their armory when determining how to respond to international situations and crises.

Policy coherence is also the capacity of governments to look beyond conventional reactions to international situations and crises—through either the military or the diplomatic or foreign aid—towards new ways. These ways would look towards epidemic or pandemic control programs that are smart enough to make broader differences; towards obtaining a seat for infectious disease control at the highest decision making-tables so that epidemic control efforts can justify themselves in the corridors of power on both health and nonhealth grounds. All of those involved need, ideally, to be in the position to describe which programs will respond to both health and nonhealth prerogatives around the world at any given time. Such presence can lead a paradigm shift towards integrated and multilevel responses that employ infectious disease response programs as tools of both altruism and enlightened self-interest.

The concept of enlightened self-interest epitomizes epidemic and pandemic responses as high politics, and its new place in the collective consciousness. Both enlightened self-interest and policy coherence are spin-offs of the wired age in which we live: by facilitating, say, public health emergency efforts to both evolve and become more connected with security, economics, conflict, and stability. This connects with the high end of the political spectrum—with prime ministers and presidents talking about communicable diseases such as HIV/AIDS, tuberculosis, and malaria. Heads of states are building more than political capital in so doing; doing more than paying lip service to health crises or enhancing their own national prestige on the world stage: leaders are recognizing that the right programs for epidemics can also address bigger pictures.

Prime ministers and presidents may thus be tuning into the revolution, and adapting to it, but the *de facto* revolution in synergistic approaches to world affairs is happening from the ground up. From groundswells as much as decisions made by statesmen: the latter's decisions, governed by the voice of the people. This voice is demonstrating that the world and its citizens now see infectious disease control efforts as going beyond pure altruism or charity: that there are tangible and vital benefits to both internationals and locals, developed and developing countries, and across the political spectrum, in true postpartisan style. Smart approaches are thus creating new rationales for global health program support and involvement—rationales so compelling that the world cannot afford not to invest. Nw rationales, approaches, and consensuses—but ones that lie on historical precedents.

Throughout modern history, there are examples of smart approaches to epidemics and pandemics that have bridged development to diplomacy, international relations, and beyond. But such examples are only *ad hoc*; they have never been structured in ways that have guaranteed optimal health and non-health performance every time. These approaches have been characterized by serendipitous, inadvertent bridges, characterized exception to unjust colonial and empire systems; dreamt up by colonial powers sending public health officers to the edge of the earth. Public health officers with duties to protect the local, native populations as much as the expat internationals; to demonstrate that colonials cared, as much as conquered (Fig. 4.5) [5].

Figure 4.5 Education and health, anyone? Without cross-sectoral collaboration and expertise at the highest levels, such efforts may struggle. HIV/AIDS education meeting, Kisarawe, Tanzania. *Picture courtesy Sebastian Kevany.*

Cold wars can be malign, and cold war foreign aid can be manipulative—but most would agree it is often better than hard power. But even cold war infectious disease response efforts mean more money for global health, whatever the underlying rationale; money that would never have been spent without ideological overtones, or enlightened self-interest. It is arguably often a better investment than arming rebels, or orchestrating *coups d'état*, or proxy wars: it is money for sick villagers at the end of dirt roads, who don't care why the internationals are there—to win their hearts and minds, or just to help—so long as they improve the health in the area.

While there have been such *ad hoc* successes in the past, what is required today is formalizing, predicting, guaranteeing dual benign effects, representing a revolution in policy planning. A new style of government, which purviews overlap is required, and where certain boundaries blur. While there will always be purviews, there is a more explicit way of doing things emerging with clearer understandings of the mechanics of interdepartmental synergies. I experienced that in Bulgaria, working on the design and delivery of local HIV/AIDS programs, where the locals identified the spin-offs and dominos from health programs that advanced the social, the cultural, the economic—as well as the postSoviet. There, they found synergies that depended on an appreciation of bigger pictures—that relied on epidemic response downstream effects being tied, not just to better health services, but also to social justice and humanitarianism, equity and integration.

As policy processes evolve and smarten, as the medical bleeds into new purviews, its previously fringe appeal broadens. Barefoot diplomacy's post-partisan appeal can transform aid skeptics into inadvertent paragons of progressive and altruistic values; and into supporters of global epidemic disease treatment and prevention programs. Lines and interests become blurred, and the appeal of epidemic responses becomes less selective, more palatable and attractive; more useful to and respected by the entirety of the political, social, and economic spectrum. Barefoot health diplomacy is thus not only for the left, but for everyone, through multilevel returns on bipartisan and post-partisan agendas. Essentially, it gives generates on different perceptions of what constitutes valuable, meaningful, investment.

But bipartisan, postpartisan support requires the integration of epidemic control or public health emergency responses and the diplomatic: a political activity, which meets the dual goals of improving health while strengthening relations among nations [6]. This then requires that global health enhances international relations, and never threatens them. It requires also that funding cutbacks or epidemiological evidence doesn't inadvertently—as a result of epidemiologists or economists operating in silos—provide infectious disease programs to only one side of a country—to one tribe, ethnicity, or political homeland.

Diplomats should maybe, then, ensure that infectious disease control doesn't inadvertently take sides—even if economists and epidemiologists suggest it as the best course. Dr. Alfred C. T. Kangolle, describes, next, the evolutions still needed within the filed of epidemiology to keep up with these demands.

The advancement of epidemiology and public health

Many public health professions may have opted to undergo the profession of public health because of other priorities. Some may have decided to study and become public health professionals because they wanted to do something else. Only a few may have gone into this field because they want to make significant contributions to public health.

What we see in any field results from long-terms conceptions in the minds of different people, which results in visions. When a vision is worked out, the ideas of how to achieve the vision becomes obvious and hence, the mission becomes clear. In so doing, multiple options occur but if the initial desire of the person is not strong, the chosen mission will fail. Persons with strong desires of joining the field are likely to have good visions and missions. Professions of such kinds will formulate the best goals, objectives, activities and interventions. in this regard, public health is the science of protecting the safety and improving the health of communities through education, policy making and research for disease, and injury prevention.

There have been some discussions that definition of public health is different for every person. In fact, every definition shows that protecting the safety and health of communities is crucial and is done by preventing the diseases and injuries. The processes used to achieve these goals are education, formulation of policy and law, and conducting research that will find the required solutions. If an individual desires to crunch numbers, conduct laboratory or field research, formulate policy, or work directly with people to help improve their health, there is a place for them in the field of public health. These professionals work around the world, addressing health problems of communities as a whole, and influence policies that affect the health of societies.

As highlighted, public health prevents diseases. Therefore, the major focus of public health should be on preventing, avoiding, inhibiting, precluding, or stopping diseases from occurring and thus decreasing overall morbidity and mortality rates. In recent years, the search for implementation of interventions that will result in stopping diseases from occurring have been either low or not given any priority. This is because there are many diverse findings. Yet spending time on outlandish interventions has resulted in the continuation of disease occurrence in poor communities.

Many authors have described epidemiology as the cornerstone of public health; this means public health is built on the foundations of epidemiology. We know that epidemiology is defined in different books as the study of the distribution and determinants of health-related states or events in specified populations, or the application of this study to control health problems. Another related definition is that epidemiology is the study of how often

(Continued)

(Continued)

diseases occur in different groups of people and why; as well as the use of such information to plan.

Some uses of epidemiology can be highlighted: epidemiological information is used to plan and evaluate strategies to prevent epidemics. Epidemiology information is also used as a guide to the management of patients in whom disease has already developed. Epidemiology information is further used for surveillance—or to monitor time trends to show which diseases are increasing or decreasing in incidence or morbidity, and which are changing in their distribution, and use such information to plan for appropriate intervention.

There is also a new discussion with regards to the definitions of public health and epidemiology. The issues under discussion are results of concern that if the appropriate definition of a term is not comprised of all important components, its use and its effects cannot be maximal. Therefore, epidemiology is also the study of the distribution and determinants of health-related states or events in specified populations, and the application of this study to eradicate or control health problems in that population. Epidemiology can also be defined as the study of how often diseases occur in different groups of people and why; and the use of such information to plan interventions that will eradicate or control diseases.

Public health prevents diseases, and preventing means avoiding, inhibiting, precluding, or stopping diseases from occurring or from causing morbidity and mortality. Then this falls in line with the application or use of epidemiology in planning for interventions to eradicate or control diseases.

Thus the discussion is now to emphasize having the world eradicate diseases and injuries in the definition of epidemiology. This new or improved definition will require more efforts in the fields of epidemiology and public health. It will need research on advanced techniques, interventions, policies and laws to be implemented for the betterment of the life of the communities and societies; it will require empowering the professions in this field, and organizing required resources. It will mean reminding all professions that eradication of certain disease(s) will not mean they will no longer be needed: our world is still having many health challenges; eradication of one disease will yield new opportunity for the professions. Chickenpox was eradicated—but the work of epidemiologists and public health personnel have never stopped.

Global visions to eradicate certain diseases should also be widening to go beyond the current focus. A few examples to be discussed and considered are: River blindness diseases—since it is known that this disease is transmitted by the insects whose life depend on the river in that community, focusing at eliminating the insects from that river will yield eradication of the disease.

(Continued)

(Continued)

Malaria has been in this world for many decades, if not millennia. Much research have been done, and most information is available. It is known that malaria parasites have their life cycles in humans and in mosquitoes: therefore, interruption of these life cycles by blocking the transmission from mosquitoes to man will result in the end of the disease.

Preventing the diseases process so that the disease will not occur, can lead into eradication or control of other diseases. For example; public health professions and epidemiologists can work with food production, food distribution, food process factories, as well as law and policy makers, to make sure foods which cause diseases in communities are not produced or distributed for use.

Another example is to make smoking and alcohol consumption illegal in all communities worldwide. Governments and all stakeholders need to know that the income (money) generated from these business has a bad impact to the health of individuals, their families, the communities and their nations. The cost incurred to care for these patients has not been well looked at. There many other diseases and conditions that, when looked at closely, can be eradicated.

In order to achieve eradication of diseases in our communities, quality epidemiology and quality public health should be the priority. Recently, we have seen more efforts that have been put into quality in health care. The results are good, and health system and health-care services are improving. It is time now to consider a similar approach of quality in public health and epidemiology; hence the new definitions of epidemiology will be useful, and will bring new opportunities to the public health and epidemiology fields.

Alfred Kangole

4.3 Innovative policy-level combinations

Within epidemic or pandemic responses, then, as in any international endeavor, the "price of peace is eternal vigilance"[1]—the vigilance of diplomatic oversight of medical programs; of avoiding potentially precarious international relations effects. One can avoid these consequences by breaking down barriers of expertise and insight—by increasing sensitivities and awareness of cultural, religious, economic, social, and political issues. Put simply, we can avoid dualities in policy, foreign, and global

---

[1] A quote originally attributed to Winston Churchill, British Prime Minister during the Second World War.

health affairs through a unifying framework of principles that govern the altruistic through the diplomatic—and vice versa.

Lack of sustainability and transferability in epidemic response program design, for example, can do more harm than good. The absence of sustainability measures fuels the fires of the skeptics, while also disappointing the locals. The skeptics, both local and international, say: "maybe you shouldn't have come here in the first place, if you plan to leave work unfinished; if you only end up breaking hearts." Skeptics, in turn, can be refuted by smarter programs: by design and delivery that mitigates such risks; by the loyalties and allegiances generated by barefoot diplomacy endeavors that both respect locals and stand the test of time.

Refuting skeptics is also consequent on programs advancing economic growth and activity for both donors and recipients [7]. Skeptics will also be refuted by changes in donor image through more diplomatic infectious disease control efforts—the international prestige that forms the centerpiece of so many nations' dreams and aspirations. The prestige and cachet of a virtuous country on the international stage, of a country respected as a global force for good; the prestige that is dependent on global health becoming both altruistic and diplomatic.

Dovetailing the medical and diplomacy will thus improve prestige, the economy, security and peace—but at a price: the price of interdigitation. The price is red tape and procedural approval: programs being vetted, not just from perspectives of medicine and bioethics, but through the lens of human rights, security, economy, and diplomacy as well. These lenses are benignly cognizant of dark, *Constant Gardener* [8]—style scenarios: lenses that prevent the health sector developing a damaging indifference towards bigger pictures. These lenses also enhance oversight and integration to ensure that infectious disease control meets diplomatic standards, as well as medical: approvals that might review the range of response interventions available, cross-checking them for adherence to both medical and holistic criteria.

Such lenses and approvals could, in theory, enhance global health's contributions to bigger pictures—but, they should also know their limits. Having such limits preserves the idealism, and nobility of epidemic response efforts in its purest, most isolated form—untainted by *realpolitik*. These limits should recognize a natural aversion in altruists to be connected with less palatable more complex considerations; limits that also ensure programs are beyond manipulation for goals that they didn't sign up for—even if, implicitly, the world has always worked that way. These are limits that keep infectious disease control beyond the reach of hidden agendas; that make the implicit, explicit.

What is needed, then, are both lenses and limits on lenses that help to govern the revolution: that calibrate the bridging of public health emergency responses and the broader aims of government and policy in a mutually beneficial way. Specific institutions may indeed already exist to address bigger pictures—conflict resolution, foreign affairs, diplomacy—that function independently of global health in those realms. Yet limits that recognize siloed expertise and stove-piped departmental independence also need to recognize past failures—that, today, no part of an engine, or a world, meaningfully operates in isolation. Good engineers, today, are also environmentalists; good football players are ambassadors; good lawyers are ethicists; and good doctors know the law. To optimize what they do, international health and medical personnel also have to be barefoot diplomats; nonhealth skills have become a *sine qua non* (Fig. 4.6).

Figure 4.6 In epidemic situations, supply chains are essential—again, requiring careful coordination between and across government departments. *Picture courtesy Sebastian Kevany.*

Lenses and limits: what maybe needs to be recognized is that the failure to consider foreign policy or international relations objectives when designing, selecting, and implementing global health programs runs the risk of creating tense and confusing dualities [9]. What might better be ensured is that the left hand knows what the right hand is doing: compartments and divisions, walls and barriers, exposed as both artificial and even destructive. In a smart world, higher expectations that medical interventions will achieve diplomatic goals beyond technical objectives will, otherwise, be thwarted by these gaps [10]. Lenses are needed that see that results consequent on bridging such gaps should be recognized and rewarded: if global health programs are only expected to advance health, then that is all they should be recognized for.

Revolutions are not always sudden and visible and dramatic: some revolutions are gradual and subtle, and are already happening inexorably, under our noses. In the United States, the revolution of governmental integration is well underway, as represented by the Office of Global health Diplomacy— [11] an organization with responsibility for taking joint consideration of infectious disease control and other foreign affairs. The creation of this organization represents smart, efficient, effective government at its best

Such offices of global health diplomacy are thus what is needed to govern smart policy; to consider what else the medical could and should do, while it goes about its primary business. To consider the trade-offs between one program and another: not just in terms of health outcomes, but in terms of the broader, dual effects of each. Offices of global health diplomacy are what is needed to evaluate the different effects of different programs on extremism, instability, terrorism: effects realized within limits—and without burning the bridges, or compromising the security privileges, that infectious disease control neutrality has earned.

An office of global health diplomacy is perhaps what is needed in Northern Nigeria, to consider the dual crises of HIV/AIDS and Al-Shabaab kidnaping schoolgirls [12]. Such an organization could optimize dual effects by combining ideas, disciplines, and players by facilitating new, innovative partnerships between diplomats, locals, and internationals—and between the medical, the governmental, and national and international security. An office like this would also ensure that politicians, multilaterals, and bilaterals all get to the same page in terms of being willing to deal with both health epidemics and kidnapings—all pulling in the same direction, at the same time.

We get these players all on the same page, n theory, only through local and international, bilateral, and multilateral, offices of global health diplomacy—through epidemic response efforts developing capacity to operate on multiple levels. But there is a need to develop capacities in people also—in barefoot diplomats, local or international, who can advance both diplomacy and health in the rarified air of high-level summits and meetings, or in the warm breeze of day-to-day field work. Epidemic response personnel taking opportunities to advance broader agendas—just as much as the environmentalist, the diplomat, can choose to advance health. Just as much as others can contribute to epidemic control or public health emergencies—navies, peace corps, departments of transportation and commerce—the increasingly diverse list of collaborators and implementers goes on and on [13].

Collaborators and implementers from different backgrounds, all working together; harmonizing to develop new mechanisms for diplomacy dovetailing with development—working for enlightened self-interest, as well as altruism. However, as our next essayist, Brian MacDomhnaill, demonstrates, this ideal is dependent on history and politics—even in the Amazon.

Um Ano Como Brasileiro

(From a diary): Well, I've been working a lot ... To gain legitimacy for the new democracy, the 1988 Brazilian constitution established a system of councils to institute public control over government. People generally don't participate much though, and in some towns government officials sit down and write up imaginary council reports and then call around to their friends to get it signed off.

I have been organizing community forums through the *Council for the Defense of the Child*, with the hope that the forums become, in effect, AGMs of the Council, and thereby get people involved—and the council up on its feet. To do this, though, we need the support of mayors; and the mayors, not liking these pesky councils, would rather not help. There are municipal elections this year and we are telling mayors that these forums are an opportunity to address a lot of people. The mayors realize that they'll have to speak a certain amount about children, but maybe don't realize that—having done it once—they'll have to do it again, and each time they'll have to defend their record in front of the people who could vote them out of office. And so we think, they might find it prudent to invest more in health and education.

Or maybe not: Amazônia is a place of mysteries; a dirt road the government sent millions to tarmac; receipts for school buses that never appeared; detailed reports on social projects that don't exist; the equally amazing spread of luxury apartment buildings, wood mills, and ranches. The basic problem in Amazônia is that all attempts to establish a sustainable economy have failed.

(Continued)

(Continued)

Trees grow to such grandeur, not because of the soil but by holding humus with their roots. When the trees are cut, the nutrients run off, and the land isn't arable. Most people just subsist with hospitals, schools etc. paid for by Brasilia (el capital). Federal resources are channeled through town councils, and thus, being the man with the money, the mayor is like a king in his municipality. And like a king, he often hasn't the least scruple in not sending an ambulance to a critically sick man if he happened to vote for the wrong party in the election.

But for all that, change is flourishing in Brazil. It is an old trick of politicians here to arrive in villages the day before an election and hand out ten or fifteen dollars to buy peoples votes. It went on again in the last election. It was noticed, though, that a lot of politicians who forked out didn't get elected; that is to say, the villagers were taking the money—and then voting for someone else! The dictatorship, after all, failed because the people wouldn't go along with it anymore. Nowadays they notice themselves to have a certain space to play with, and maybe even to speak up in our little forum.

Back to history: agriculture having failed, the colonists in Belém, finding themselves in vast dimensions of useless trees and water, poor and frustrated more than anywhere else in Brazil, turned to the exploitation of people. Boats were sent upriver to hunt indigenous people, leading to a personal and social de-structuration. The treatment was so severe here that slaves were sent to Amazônia as a punishment. So, I suppose, it's not surprising that people in Belém are grumpy and badly mannered.

For a long time, and even now, I found it hard to associate with the place. As time has gone on, and my life has become more normal, I have come to take for granted certain things. Belém is known as *Cidade das Mangueiras* because, mango trees reach up and mat across all the avenues of the city. Every morning, there is a big blue sky, but as it gets warmer, clouds form, darken and then it lashes rain. After the rain, the city is steamy. People leave their offices, or come from their apartments to eat on street corners, play cards, go to church and the like. You always see someone going by with a guitar, or other instrument. At night, the bars are full.

There aren't many bars for my sort but one, *Café Imaginário*, is in a half wrecked old colonial house. The doors and windows wide open, people arrive after midnight; it's different to an Irish pub—or maybe like an Irish pub in the old days where people sit down, half listening to the music, half talking to one another, except here it goes on until morning, and even then they are still playing on base and sax.

All night, boats have been coming in from the islands stacking up their produce. Going home in the morning it is warm; harbor is jammed with

(Continued)

(Continued)

people selling and buying. I suppose I'll miss all that. They say that the European way doesn't keep in the tropics; that you either live as a Brazilian, or go demented. But, for me, I didn't adjust to the life because I don't admire it. It has its moments, but it's not living in peace, or at least not for me. Still, I'll miss the swagger and pride of the *brasileira*.

I was planning to go back to Ireland, but I stayed on to pilot the forum. I was thinking of staying on again until December but then, after that, there probably would be something else again, and I could easily end up spending five or six years in Latin America. So I'm saying goodbye to Brazil, and to UNICEF, and the adventure that was living here. I wonder if the place will ever change—and what will be left if it doesn't.

Brian MacDomhnaill

4.4 Dynamic field-level combinations

But new mechanisms are needed that bridge gaps, create synergies—that make the whole greater than the sum of its parts. Mechanisms are needed to build lines of communication and consultation between presidents, prime ministers, and global health: mechanisms to expand the responsibilities of barefoot diplomacy offices to advise on the potential contributions of epidemic control efforts to world stability, conflict resolution, and international relations—in both emergencies, and over longer terms. With new mechanisms, we can articulate and optimize latent, untapped contributions of infectious disease control to the bigger picture and also respond to national and international crises and concerns. These mechanisms will facilitate inputs and suggestions, contributions and ideas, about the pressing tragedies of the global community not just from politicians and statesmen, but from barefoot health diplomats as well.

Global health's potential contributions are thus dependent on developing the capacity for broader perspectives—dependent on the capacity of barefoot health diplomats to make holistic recommendations on how smarter programs should be designed, in terms of bigger pictures. But to do this, such diplomats need an understanding of the challenges and feasibilities of different response scenarios, on both *ad hoc* and routine bases, to the world's immediate and long-term health and nonhealth concerns.

I saw the need for such capacity in Papua New Guinea: amongst Australians and Papuans, amongst diplomats, doctors, and nurses. They understood the connections between local problems with tuberculosis, malaria, HIV/AIDS epidemics, their associated international responses, and international health security: understood also the need to tune into links between migrants and refugees; contagious disease drifting the fifty miles across the Darwin Strait; links between HIV/AIDS and tuberculosis in the Western Highlands of Papua New Guinea; and economic migration. All of these aspects were spiraling with the bigger picture—the discovery of oil and gas resources—and all were affecting health and diplomacy, the economy, and international relations.

In Port Moresby, the capital city, the lines between such purviews were rapidly becoming blurred. International hospital ships blurred distinctions between epidemic response programs and trade deals, politics, and economics: the ships were benign forces—yet inevitably, implicitly, in local eyes they were benignly linked to nonhealth agendas. They were implicitly linked to international goodwill and relations, in the same way that Australian epidemic response programs in the Western Highlands implicitly overlapped with bigger pictures: the conflation of epidemic control and natural resource extraction.

Other examples from my own experience abound: explicit, coherent global health and diplomacy agendas exist from Australia to Ireland. From Australia's smart approaches to epidemic control in Papua New Guinea to Ireland's international development policies that now link epidemic or pandemic responses with trade, international relations, and international image: [14] policies and programs that advocate for development, best achieved when all government policies and actions complement each other—[15] through a whole-of-government approach [15]. Breaking down barriers and challenging artificial distinctions and antiquated traditions can now ensure that diplomacy enhances the humanitarian, and vice versa (Fig. 4.7).

Notably, barefoot diplomacy approaches around the world can also leverage donor contributions to multilateral organizations to ensure harmony with national interests: these approaches ensure that donor contributions, for which governments are feted and admired, also address diplomacy, conflict resolution, or post-colonial angst. France insists that specific amounts of its donations to the Global Fund to Fight AIDS, Tuberculosis and Mamaria are assigned to Francophone West Africa: the synergies that result, in turn, both guarantee and augment global health investments—investments that wouldn't even occur without their increasingly explicit, downstream returns.

These overlaps and synergies also justify a recalculation of the investment ratio between epidemic control and military force, because they also advance

Figure 4.7 In rural Papua New Guinea, Australian-led infectious disease control programs have supplanted the old "colonial officer" system. *Picture courtesy Sebastian Kevany.*

the strategic. Avoiding hard power through barefoot diplomacy is a basis for *avant garde* collaborations, institutions, and approaches: overlaps in such investments can become the basis for the foreign policy and security strategies of a European Union built on a last-resort perception of the use of military force [4]. Infectious disease control provides alternatives to hard power as much as it harmonizes national approaches: a preference to avoid ballistic interventions does not have to mean meek foreign policy Harmonized approaches to infectious disease control programs across member states are increasingly based on common sets of beliefs, practices and principles: these principles also advance world peace, cooperation, and stability; can enable epidemic responses to act as a new option, or an innovative tool, in response to world issues and crises.

Many evolutions in global health, epidemic control, and diplomacy approaches, both in government and in foreign policy, are emerging before our eyes. In the United States, the United States Agency for International Development, the President's Emergency Plan for AIDS Relief, and the State Department are working more closely together now than they ever have before [16]. In the United Kingdom, the Department for International

Development is increasingly inseparable from Whitehall [17]. Around the world, the links between health and diplomacy, are requiring governments to re-examine resources necessary to achieve mutual objectives: [2] smart revolutions that enable the health of the world's poorest people as much as international security can be advanced by every means, every department, every policy—to ensure that health advances peace and stability; that health gains do not come at the cost of international incidents.

A revolution is now perhaps underway that ensures responses to epidemics or public health emergencies tunes in to bigger pictures—and vice versa. But understanding bigger pictures and optimizing synergies requires politicians and diplomats, statesmen and women, to understand the implications of epidemics on conflict, human rights, and the environment: governments should maybe try to understand the impacts of different program choices in different places, at different times. To deny those effects seems like defying the inevitable: to whistle past the graveyard. In the 21st century nothing operates in isolation—one act inevitably has effects on others (Fig. 4.8).

Figure 4.8 Governance doesn't always take place in stuffy boardrooms. Here, locals and internationals exchange collaborative ideas over high tea in Mutotko, Zimbabwe. *Picture courtesy Sebastian Kevany.*

4.5 Key messages

- Traditional deprecations between governance, such as between foreign policy and global health concerns, are no longer relevant.
- Instead, a much greater level of interdigitation and interdepartmental involvement is taking place.
- This requires personnel versed in different specialities, and being able to communicate with experts in other fields.
- In turn, this helps to identify synergies between departments, such as environmental goals and epidemic control.
- Such sector-wide, policy coherence or whole-of-government approaches free up resources use where they are most needed (Fig. 4.9).

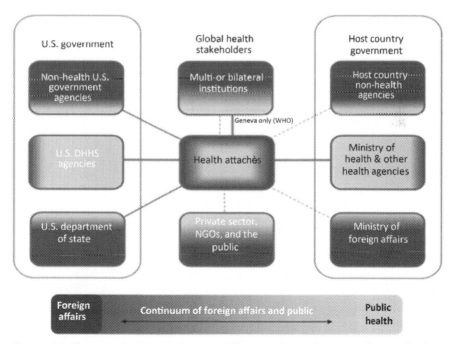

Figure 4.9 The complexities and many different military, foreign policy, and other bodies required for the combination of diplomacy, global health, and infectious disease control. *Courtesy Matthew Brown.*

References

[1] Kevany S, Fleischer T, Benatar S. Improving resource allocation decisions for health and HIV programmes in South Africa: bioethical, cost-effectiveness and health diplomacy considerations. Glob Public Health 2013;8(5):570—87.

[2] Royal Institute for International Affairs. Global health diplomacy: a way forward in international affairs. Chatham House; 2010.

[3] Walensky RP, Kuritzkes DR. The impact of the President's Emergency Plan for AIDS Relief (PEPfAR) beyond HIV and why it remains essential. Clin Infect Dis 2010;50(2):272—5.

[4] Kevany S. Global health diplomacy: a 'Deus ex Machina' for international development and relations. Int J Health Policy Manag 2014;3(2):111—12.

[5] Paxman J. Empire. London: Picador; 2010.

[6] Novotny T, Adams V. Global health diplomacy—A call for a new field of teaching and research. San Francisco Med 2007;80(3):22—3.

[7] Kevany S, Murima O, Singh B, Hlubinka B, Kulich M, Morin S, et al. Socioeconomic status and health care utilization in rural Zimbabwe: findings from project accept (HPTN 043). J Public Health Afr 2012;3(e13):46—51.

[8] Le Carre J. The constant gardener. London: Penguin; 2009.

[9] Center for Strategic and International Studies. Final report of the CSIS commission on smart global health policy; 2010.

[10] Katz R, Kornblet S, Arnold G, Lief E, Fischer J. Defining health diplomacy: changing demands in the era of globalization. Milbank Q 2011;89:503—23.

[11] Kevany S. James Bond and global health diplomacy. Int J Health Policy Manag 2015;4(x):1—4 Editorial commentary.

[12] Kevany S. Global health engagement in diplomacy, intelligence and counterterrorism: a system of standards. J Polic Intell Count Terror 2016;11(1):84—92.

[13] Vanderwagen W. Health diplomacy: winning hearts and minds through the use of health interventions. MilMed 2006;171:3—4.

[14] Kevany S, Brugha R. Irish aid and diplomacy in the Twenty-first century: optimizing enlightened self-interest, supranational priorities and foreign policy impact. Ir Stud Int Aff 2015;26:207—25.

[15] Government of Ireland. One world, one future: Ireland's policy for international development. Dublin; 2013.

[16] Kevany S. New roles for global health: diplomatic, security, and foreign policy responsiveness. Lancet Glob Health 2016;4(2):e83—4.

[17] Lords Select Committee. Does UK aid make the right difference?; 2011.

CHAPTER FIVE

Monitoring and evaluation: capturing barefoot effects

Abstract

The role of monitoring and evaluation has been a key driver in changing perceptions of quality and value of global health programs. However, the focus on a limited range of outcomes may result in those results that are not considered, or other unintended consequences, being missed. As a result, health programs that may function highly on a diplomatic level can be totally unrecognized in assessments of resource allocation or technical efficiency. In this chapter, innovative ways of capturing the nonhealth effects of a global health program in realms such as the environment and conflict resolution are presented, which should be taken in to consideration in combination with their clinical and medical effects.

5.1 The pressures of targets

In the world of global health—and any other discipline, for that matter—monitoring and evaluation is never going to be particularly stylish, fashionable, or inspirational (Fig. 5.1). Monitoring is the means of tracking performance and keeping a finger on the pulse; evaluating is the means of looking for changes in results. To do both together, you need to put an ear to the tracks; two ears, operating in two very different ways. A system of checks and balances that is often just too numeric—assigning numbers, trying to quantify, and often striking the wrong note.

Yet monitoring and evaluating is a means of investigation. It means making sure the money is going where people say it is going; that people are doing what they say they are doing. It is ensuring that the money is getting the correct number of people being treated for malaria; the correct number of people tested for new infections. It is all fair and benign—as long as you believe that everything that matters can be and should be measured, enumerated, or quantified. If not—if everything good cannot always be measured in life; if you cannot see the wood for the trees, or if the numbers hide more than they reveal—then monitoring and evaluating can be a double-edged sword.

Sometimes, all the distant internationals see are those results, those numbers. It is often those numbers which are noticed and discussed in houses of

Barefoot Global Health Diplomacy.
DOI: https://doi.org/10.1016/B978-0-12-818681-7.00008-X

Organization of U.S. Global Health Efforts

Figure 5.1 Monitoring and evaluation results for epidemics go all the way to the top.

parliament, at kitchen tables over dinner. "Ten million HIV/AIDS patients on antiretrovirals"—that information that comes from the field; from all the way down to the patient or the nurse, way out in the bush. It comes from a swamp, a back alleyway, a battlefield, a desert: all the way from names in a ragged patient register. It involves matching details and diagnoses, checking patient files, and getting the numbers right.

But the numbers sometimes hide as much as they reveal. One example is the measuring of bed net distribution for malaria—an intervention to ensure that people do not get bitten when they are sleeping. The numbers may say three hundred nets distributed in a week—but what does that mean? Did anyone use the net after you gave it to them? Often—by lakes or seas or rivers—the malaria nets end up as fishing nets. Sometimes, the nets are never even taken out of their packets. They are never used because locals do not feel comfortable with them; because they are an alien presence in the home.

Measurement and performance pressures create their own problems, as well. Livelihoods depend on performance, and often the numbers suggest that the end justifies the means. Locals may say: "We will do whatever we can, whatever we have to do, to meet the targets." Locals are often told by the internationals to "come in to the clinic now, and get tested." Come in right away, is the message, because the implicit underlying meaning is that people have to hit their targets, by the end of the month. Coercive tensions are an ethical dilemma, but they are also a reality. Managing the reality thus requires vigilance, but it also requires sensitivity and respect: respect for

Figure 5.2 In South Sudan, monitoring and evaluation efforts were inherently collaborative and team-building exercises. *Picture courtesy Sebastian Kevany.*

patients and for providers, both. Those moments, when targets and indicators create malign pressures, should maybe be acknowledged: the flash points have to be recognized, when monitoring and evaluation goes up against cooperation, diplomacy, and sensitivity [1] (Fig. 5.2).

Monitoring and evaluation can also create pressures on the smallest health centers, which lie way out on the edge—running on a shoestring, with all hands on deck. This creates more responsibilities for the local nurse—on whoever is there—who has to run the numbers. On top of the pressures of running the clinic day-to-day, they also have to fill out forms at the end of the long shift, the long week. They have the added work of filling out registers, detailing the characteristics of what they have seen—the minutiae of what they have done.

These added pressures suggest the importance of both locals and internationals tuning in to new paradigms, new styles, for monitoring and evaluation. Tuning in to new questions on the effects of infectious disease control on the relationships between the local and the international: the internationals too often coming in and pushing for information, transparency, and results with no time for finesse, acknowledgment of ripple effects, or details.

So the current groundswell—to monitor, to evaluate—also needs to evolve and adapt. Evolve to capture broader effects such as peace keeping, nation building, conflict resolution, international relations, or world security; adapt to create better relationships between locals and internationals. How or if that can happen is key to renaissance, avant-garde, epidemic control efforts. It leads to questions about how programs can be designed optimally; how systems of monitoring and evaluation can capture magical results. Such new questions represent a quantum leap, revolving around how we measure achievement in the 21st century. A shift in what we consider constituting success; a shift in the evolution of benchmarks against which our endeavors are measured.

We thus require new benchmarks to monitor and evaluate the ripple effects; we need new measures for proving these programs are effective and to generate new systems that guarantee these effective results every time. Systems that—in capturing the ephemeral—also capture a rationale for support; that interventions can be done in a way that advance health and diplomacy, health and world stability, health and the environment—at the same time, every time [2].

But can the downstream, domino effects of global health could be measured? Could you measure such effects any more than you can quantify the effect of weather on mood; could you put numbers on hearts and minds—on sustainability? Is it possible to build a capacity in infectious disease control—in anything—to demonstrate effects beyond a gut feeling, a sixth sense, a hunch? Can you make it manifest—beyond just the intuition, the instinct, that you are operating benignly on more than one level at once? Dr. Dianna Kane, in our next essay, provides an example of how numbers can miss intangibles.

Medic Mobile—team diversity and global health diplomacy

I joined Medic Mobile in early 2012. We were fewer than 10 people, a fledging organization exploring the use of digital technology tools to improve healthcare delivery in hard-to-reach communities around the world. A largely North American staff at the time, many of us had worked abroad in the nonprofit sector, and knew that to effectively design locally appropriate products, as well as to scale and sustain them, it was critical to have a meaningful connection to the countries in which we worked. From the beginning, we knew that building Medic Mobile would be about more than just shipping software; it would be about building long-term partnerships and cultivating local expertise in what was then a very new field.

Now, in early 2019, we are a team of close to 100 people. This includes well over a dozen nationalities in more than 25 cities spanning East and West Africa, South Asia, Europe, North America, and Australasia. We have three physical offices

(Continued)

(Continued)

in Africa, one in Asia, and one in the US, and remote teammates working from many other places. We have intentionally established diverse teams that span regional locations, and as a leader of a cross-regional team, I get to experience this diversity daily. It is a marvel that 21st century technology, with all of its challenges, has allowed us to connect and collaborate in this beautiful way.

It has been a priority of mine for our design team, comprised of 14-people in five countries, to maintain a strong unified identity. We have a standing call every other week in which we share user insights, discuss design methods, and collaborate on new features and workflows. In the moments before we come together, two of us are waking up in Portland and San Francisco, another is having afternoon coffee in Dakar, several are cooking dinner or traveling home by taxi in Nairobi and Kampala, and others are winding down for the evening in Kathmandu. Through consistent calls and in-person meetups, our teammates have gotten to know each other personally. As we start up our meeting, there is often friendly chit-chat, questions about children, weekend plans, who is eating what for dinner: I feel like I am in the company of friends, yet I know how uncommon it is to have a friend group this diverse.

We also routinely host in-person meetups of teammates, but the ultimate example of this was our first Medic Mobile full-team meetup at a beachside resort in Kenya. On an afternoon break from work sessions, I watched as Nepali, Senegalese, Ugandan, British, Ethiopian, Kiwi, and Kenyan teammates (to name a few), played a vigorous game of water basketball in the hotel pool, cheering each other on and offering high-fives. Later, we took turns performing karaoke songs from our countries and we danced into the night. The next day, everyone was at their desks, ready to solve some of the world's biggest problems in healthcare.

It is also important to honor the sacrifices required to make this work. The context behind the moments leading up to our design calls contains a range of challenging situations. One person may be sick while traveling, another may be tired from getting up early for calls all week, someone may be missing dinner with family, and another may be frustrated with a poor internet connection after a long day in the field. As designers, we practice empathy as a profession—and this serves us well in our dealings with teammates, too. We have developed a sixth sense for when we need to see each other in-person, or for when the meeting format needs to change to better include someone who is consistently quiet, or has become disengaged. Taking time to understand and care for each other provides a greater depth, connection, and respect for one another: we are not purely professional collaborators, but whole individuals with stories, concerns, and challenges. This strengthens our commitment both to the work we are doing, and to our support for each other in doing it.

(Continued)

(Continued)

It is humbling to see such diversity come together—amidst considerable sacrifice and challenges—to solve problems that unite us. Josh Nesbit, the CEO at Medic Mobile, often leads an exercise where he asks an audience to draw a circle around the people they care about. It is common for people to start with their family and community. When encouraged, this might expand to include your home country. Our mission is to continue extending this circle to everyone in the world, to ensure that those who are most overlooked and hardest to reach are included. In addition to this mission, I can see that we have accomplished something else that is deeply important and needed in the world: it is a rare occasion for 100 people from such varied backgrounds to know and respect one another in such a multi-faceted way. May this be a model of what is possible: if our schools, workplaces, governments, and social lives operated in this way, our world would be a very different place (Fig. 5.3).

Figure 5.3 The Medic Mobile Team. *Courtesy Dianna Kane.*

Dianna Kane

5.2 Capturing downstream benefits

Monitoring and evaluation already does so much that is hidden—by developing human capacity and skills. Skills used to record the number of patients seen, complete forms, and to send forms from distant health clinics to district offices—from the district to the capital, to the ministry—ever upwards. Skills are honed in reporting flow and in the building of partnerships and collaborations between locals and internationals. Skills are developed in harmonizing: ensuring efforts, projects, and initiatives are using the same measures that make intervention results comparable across countries and villages, populations and communities.

Monitoring and evaluation also produces indicators, targets, values, performance descriptions. Indicators such as bed-net distribution, the number of HIV/AIDS patients taking antiretroviral drugs, or the number of people screened for tuberculosis are all measured and produced but how do we measure the real truth of what is often so clumsy and proxy and subjective? Do you count the number who sign up for antiretroviral treatment, or the number who take their drugs; the number who take their drugs sometimes, or the number who consistently take their drugs and adhere to the regimen so that the drugs are effective? Contemporary efforts are limited—but they are also illuminating, enlightening, and informing. They are imperfect best guesses, in an imperfect world.

Monitoring and evaluation also advances transparency; it is part of the anticorruption effort. In the old days, there were too many unknowns: too few numbers—true numbers—were sent back to donors—to taxpayers, politicians, voters; to the man or woman in the street. When they did exist, the numbers were too vague and the situations were too dodgy: situations in which money is sent and—voila—the minister for health might have a new Ferrari. Of course, the graft and brown envelopes and greasing wheels is happening everywhere, all the time, in all walks of life. It is a necessary evil, maybe; what makes the world go around. But it is not always a good thing: not always right. Not right, in particular, when the money could be spent on medicine instead.

The right response lies instead, maybe, in is increasing efforts to track, verify, and build up a pride in numbers—in productivity, from the nurse in the distant rural health clinic, all the way to the top. These efforts instill a sense of accomplishment in meeting targets; building a belief in the value and virtue of reporting achievements. Efforts to replace laissez-faire approaches are encouraging locals and internationals to want to get the reports right: to want accurate figures, as well as to want to meet the targets.

And monitoring and evaluation creates health security, as well. It provides the low-down on where the epidemics and diseases are: by region, country, population, town, or village. Monitoring and evaluation is thus a form of benign surveillance; reporting on numbers of tuberculosis cases helps to explain, define, investigate, control. It helps the internationals to work with the locals to understand what is going right or wrong: to see the broader trends, spot the gaps, triangulate impressions, build up a history—a picture of the area. Monitoring and evaluation is, today, research and learning, induction and deduction, academia, Sherlock Holmes, and discovery all mixed into one [3] (Fig. 5.4).

Ultimately, then, monitoring and evaluation tries to make sure that everything that has been done is captured and recognized. It is there to ensure that locals and internationals see results: that they understand what is going down, what has been done, and what has been achieved. It is meant to show that there are results; measures of where money has gone. This is good for the internationals, and good for the locals: recipients can see that the West, the North—the "haves"—are definitely, verifiably helping.

Even with all that, there is something missing: the smart edge. It is not just what you do, but the way you do it that gets results—results of a different kind, invisible results. The clinics and health centers I am talking about are often way out, often miles from anywhere; so the local health professionals, putting in shifts for the good of the village, do not have a lot of contact with the internationals. The locals do not get a lot of visitors—they often do not get a lot of exposure to different countries, cultures, or styles. Not until the internationals rock in with the trucks, diving deep into performance, standards, routines, and lives; diving in to local pride and dignity, as well [4].

In those situations, monitoring and evaluation requires an edge of diplomacy, and of international relations, as well. It demands efforts to be smart: to be aware of bigger pictures, reflected in ways of challenging and questioning. It is reflected also in the questions asked; in ways of looking at what the locals have done; reflected in the style in which the monitoring and evaluation is conducted.

For sure, monitoring and evaluation is about anticorruption and local performance and epidemic control—but it is also about international performance. About international style: because it is easy, too easy, for internationals to be heavy-handed: to cultivate adversarial auditor, accountant, or investigator

Figure 5.4 Can't see the wood for the trees? In the South Pacific, collateral benefits of infection control programs such as nation-building or rural development may not always be captured by the numbers. *Picture courtesy Sebastian Kevany.*

mentalities. Too easy, sometimes, for international visitors to let the style go to their heads—to mistakenly think that is what it is all about. It is easy to forget the need for gentleness, for finesse—to miss the wood for the trees. To forget roles as envoy, as representative—as communicator, or even as an entertainer.

It is thus important to be able to give a wink and a nod—to sit and listen to problems and frustrations, achievements and dreams. It is important, essentially, to try to be cool and to stay cool; because, if there is a problem—if malaria bed nets could not be distributed in the rainy season, or if the epidemic has surged in a certain population group—ninety-nine times out of a hundred, there is a good reason for it.

So, style matters, style counts. It counts also because thinking in terms of numbers, saying there is no room for comment on the scorecard, does not always jive with the reality: with the renaissance, the enlightened, the holistic. So style matters—but, often, the internationals lack the capacity, awareness, and mindset to realize this and adapt their style. They can sometimes lack the field of vision that embraces the downstream and the domino effects of their roles, so they end up missing the beauty and the artistry of what has been achieved in adversity, instead looking at narrow results from every angle. Dr. Micele Rubbini demonstrates, in our next essay, how even the most straightforward medical effort can go far beyond the quantitative.

The Namibia Project: a challenge and a proposal for sub-Saharan Africa

Global health diplomacy (GHD) is a fascinating and current topic that has involved many researchers, scholars, intellectuals and stakeholders all over the world. The common aim has been that of laying the theoretical and/or programmatic bases for an intervention, structured as much as possible in favor of a uniform and global vision of international relations based on the health of people, their protection, and the implementation of their safeguard. These relationships have been, from time to time, understood as a way of spreading cooperation between governments, such as defending interests and expectations common to populations in related areas, or as activities of associations, foundations, universities or supranational institutions aimed at raising awareness of health issues to the degree that they become central, founding elements of the relations themselves.

The scientific, as well as sectorial literature, still provides valid arguments to support the role that the GHD should have in the area of international policies—both of individual countries, and of the institutions that group them together on a worldwide scale. Alongside this substantial scientific production, we have witnessed, with increasing frequency since the end of the last century, the production of numerous documents and deliberations by international organizations of various kinds—and at various levels.

(Continued)

(Continued)

The majority of these deliberations and documents aim to outline and support guidelines and actions based on the assumption that health is an inalienable right of every inhabitant of the planet, and its preservation and implementation are the essential tools for every possible form of future development. From the fight against the diffusion of transmissible diseases to malnutrition, violence and traumas of war, GHD is rapidly and ever more convincingly becoming the tool to deal, on an international strategic level, with problems deriving even from the availability of food and water- their production, distribution and conservation as well as being the most appropriate instrument to combat inequality, poverty and violence of all kinds.

In this sense, the theoretical development and the principles and meanings that GHD can assume within various types of relationships or international collaborations continues to be enriched with contributors and actors. On a purely scientific level, contributions have been made that have deepened or widened the meaning that must be given to the GHD by broadening its definition, or redefining the areas of intervention, as well as better specifying the areas of research and the ways in which they are conducted.

New operational and analysis tools useful for making decisions have been developed and presented for evaluation to the institutional stakeholders or organizations who, for various reasons, move within this area. Yet one aspect that is still struggling to gain a foothold—and therefore risking the relegation of the entire issue that makes up the cultural area of GHD to the level of an exercise of good will, all in all inconclusive—is the direct, concrete, operational action of the governments, both as individuals and as groups organized within international forums.

In order for GHD to truly establish itself as a force capable of changing equilibrium and above all the state of relations and areas of action—as well as the living conditions of people—it is necessary that governments first-hand adopt the principles; develop the channels of intervention; and affirm its value as an instrument of overall development, both economically and socially, and as a guideline for international policies. GHD thus has the potential, if adopted as a tool for regulating relations between countries, to transform the individual and collective resources that the development of epidemic conditions can produce, in to a supranational asset that can be used to resolve conflicts and develop new opportunities to improve quality of life of entire areas of the world. The project we are working on is a modest example.

The University of Ferrara, Italy, and the NUST (Namibia University of Science and Technology) and other international partners began a collaboration with the aim of defining a curriculum for a new Bachelor's Degree for Nurses and Midwifery and the simultaneous construction of a new Campus in

(Continued)

(Continued)

Outapi, in the Omusati Region (Fig. 5.5). The project developed through three objectives:

1. Create a new workforce professionally trained to address the needs expressed in the Namibia National Health Policy Framework 2010–20: reducing the incidence of transmissible diseases such as HIV/AIDS, malaria, tuberculosis, but also fighting the incidence of noncommunicable diseases such as metabolic syndrome (hypertension, diabetes, obesity), tumours, or cardio vascular diseases.

2. Extending the reorganization of the care and research network. The new professional skills will have as a place of action not only hospitals, but also health assistance activities on the territory. The availability of a greater quantity and quality of health work force also aims to better address the health and hygiene problems, not only related to individuals in terms of prophylaxis and personal hygiene, but also to extend and implement the ability to make strategic decisions and operating on issues such as the production, conservation, distribution and quality of food—as well as access to drinking water, and the water distribution network.

3. Improving health conditions and quality of life of the populations of sub-Saharan Africa through the involvement of neighboring countries whose health plans evince a similar situation—both in terms of territorial distribution of populations, and incidence of communicable and noncommunicable diseases.

Figure 5.5 The laying of the foundation stone of the new Campus at Outapi (Namibia) with representatives of the University of Ferrara and NUST, the administrative political authorities and the Governor of the Region (2017).

(Continued)

(Continued)

The progression of the objectives is managed by three different actors: the first from the universities involved; the second from direct participation in the project by the Government of Namibia; the third from the collaboration between sub-Saharan governments. The three levels of intervention have in turn the purpose of:

1. Promoting transfer of knowledge and technologies with the aim of creating an independent, professionally skilled local workforce capable of managing the subsequent implementation of the project;

2. Supporting the Government of Namibia to adopt a territorial reorganization of health facilities, while considering also the transport network available, the level of access to treatment, hygienic conditions, availability of water sources, and communication networks.

3. Creating a consortium of countries in the sub-Saharan area united by similar health problems, with the aim of overcoming fictitious territorial divisions and promoting support relations in the development of professionalism, recognized by all the countries of the area concerned, with exchanges of staff and the creation of common databases that can be used both for research purposes and for improving the quality of health services.

The resources needed to support this project will be sought both in private sectors and, above all, via collaboration between more economically developed countries with medium and lower economically developed countries. Countries with a lower level of economic development will benefit via better quality of the living conditions of their populations and a concrete support of their development plans, as well as the possibility of main role in their own development.

For countries with a more developed economic status, the benefit could be in terms of epidemic control, international relations, and a form of reduction of the national public debt equal to the amount of investments made to support these forms of collaboration; the amount could be agreed at an international level, and managed by the bodies which control and govern the debt itself. The sharing of resources made available by individual governments with the economic and financial organizations that interact with them would therefore be an effective strategy for actions of real support for global improvements of living conditions of individuals: the truly real and irreplaceable drivers of development and well-being [5] (Fig. 5.5).

Michele Rubbini

5.3 The dangers of performance mindsets

Performance pressures in infectious disease control sometimes can therefore result in good things—but, equally, result in tricky episodes. One difficult element is focused on narrow visions—on enumeration and numbers, which eclipse all the other issues. It is the Achilles Heel of what the other shortcomings of contemporary approaches lead up to: the numbers often miss the magic—the downstream effects, as seen by local communities. But is it humanly possible to evaluate in ways that takes the hidden and the downstream effects into account: to capture the ephemeral? Is there a way to break out of the machine, out of utilitarian strictures; to loosen, broaden, to inform beyond numbers? Is there a way to reveal hidden dimensions—dimensions of the where, when, how, and why of global health that has contributed to abstract ideas such as building networks—creating partnerships, or developing senses of identity and responsibility? [6].

These ideas allow epidemic or pandemic responses to advance peacekeeping or nation building, resolving conflict, or perhaps change perceptions of locals and internationals. They also allows global health to advance international relations by addressing racial, ethnic, tribal, environmental issues—or at least, doing them no harm. If these concerns were in the past sometimes addressed through infectious disease control, what were the elements—the essential aspects—that gave the programs concerned these ostensibly mystical powers? Why did some programs have it—that edge, that je ne sais quoi, that x-factor—and other programs did not?

Some said it could not be done; not to bother; that it was a waste of time. Others said that no one, in any walk of life, could ever get a grip on the downstream effects. By the very nature the special projects and programs and responses are ephemeral, mercurial; immeasurable. Unquantifiable: in the absence of information on broader outcomes of health programs, we have failed to reach any significant conclusions about the collateral results of humanitarianism.

The absence was in the rubric; in looking past conventional measures of success. Just like the individual, knowing that a life cannot be judged based only on conventional measures of success. The ideas resonated in the real value of the lives people live, the choices we make, and the things we do—value untapped and disregarded, because it could not be measured. Because, most of the time, beyond those values people just do not want to know: it is TMI, and ends up complicating decisions, choices, reactions.

Complicating, maybe; but, in my view, complicating is better than over-simplifying—because, as they say, the truth is seldom pure and never simple. Simplification is restrictive: it is only looking at one corner of the picture, or only listening to one part of a song. Simplification means seeing shadows, not reality, because—even if narrow results perform, and even if a program looks good on the surface—when you drill down, and when you look at every part of it from every angle, it gets harder to say it is an unqualified success on every level. In the same way, the number of bed nets you distribute is not going to inform you about whether or not internationals are connecting with the locals—nor will it tell whether local cultures are embraced in a program or ignored; about whether the programs themselves are transient or permanent; about whether the environment, society, economy, or security is advanced or hindered by the program; or about whether more than just health has been gained [7] (Fig. 5.6).

Measuring the domino effects, and capturing unintended consequences flows only from the field work. Flows from the exposure and the experiences, and the imperative to innovate. To try to consider not the standard, but the nonstandard; not just the ordinary, but also the extraordinary. By optimizing the humanitarian, diplomatic, social, environmental, as well as the epidemic control element of a program can we thus somehow capture the intangible, unquantifiable, the immeasurable as well?

If you cannot measure diplomacy—cannot measure good vibes—maybe what can be done is to look at measurable things in different ways. New, gold standards, distilled from the programs that clicked: the ones that were designed and delivered, in the right way—in the right place, at the right time. One example, for me, was in the Middle East, in Jordan—with TB programs that the locals liked, and that functioned and operated on different levels. The programs that achieved the diplomatic and epidemiological results, without either coming at the cost of the other. This achieved not only narrow targets but something bigger, wider, and transcendent as well; via their location on the border with Iraq, something special and intangible, that had no name—but that we would be lost without (Fig. 5.7).

Different measures reflect different values—in global health, as in anything. Different people value different things: however, some values are universal. And people chasing value is what makes the world go around; what builds bridges, both literal and metaphorical. An engineer values a safe and strong bridge, while an environmentalist values a green bridge; a sociologist

Figure 5.6 In White Nile Province, Northern Sudan, malaria bed net monitoring visits had to take place at night I order to see if materials were in use. This required careful negotiations with local households. *Picture courtesy Sebastian Kevany.*

Figure 5.7 Jordanian tuberculosis programs on the border with Iraq included security, medical, governance and diplomacy experts. *Picture courtesy Sebastian Kevany.*

values a bridge that brings communities together, while an architect, or an aesthete, values a beautiful bridge. A poet values a bridge that stands for something; a diplomat prefers a bridge that reconciles. With structures—in fact, with anything—achieving enlightenment is therefore perhaps consequent on hitting as many benign values at once as you can [8].

The beau ideal of renaissance global health, including its monitoring and evaluation, is trying to hit these multiple values. Diplomatic values have to be taken into account, while environmental values may also weave in. Because who wants to see medical waste wrappers all over the ground of an African village? Instead, with the integration of values as part of a program, the locals and internationals see, as if for the first time, parts of the picture they had never seen before: the collateral, the knock-on, the downstream.

Monitoring and evaluation, along with the pressure to perform, were therefore perhaps limited in the past—by not revealing the whole picture; not playing the whole song. They were limited, in the 21st century, by 20th century mistakes: by McNamara Fallacies (see Chapter 6) in a new guise. They were limited by their indicators, the monitoring and evaluation, telling you that—yes, you are winning the war on infectious disease. All the indicators, all the numbers, can sometimes say the same wrong thing: the narrow measures of success indicate that everything is working, because numbers never lie; because if you cannot measure it, it does not exist. On the ground, in reality however, all those numbers are meaningless, misleading, futile— they are disinformation. In other words, to have a party and measure its success or value based on the number of people there, on the number of presents, or on the price of the gifts; on such narrow metrics, it might have been a success—but does that mean it was a good party?

But there are limits, seemingly insurmountable limits, on how the intangibles that make the difference between good numbers and good reality can be measured and guaranteed. There are limits, in epidemic responses, on measures which you can take to policymakers—to the man in the street, as much as to politicians—and prove that you are doing invaluable work, and getting hidden results. These seemingly insurmountable limits can only be overcome by working out questions, considerations, or systems that connect the broader values of international relations, diplomacy, the environment, world security, and other global goods, and concerns with infectious disease control: systems that ensure all these values are advanced in concert, without sacrificing the quality of any.

Overcoming limits requires a revolution in style and perspectives on performance that is holistic, inclusive, all-encompassing. It will require systems that over-reliance on the quantitative; that instead reveal every inch of the picture, every word of the song. That will reveal whether global health programs are diplomatically, humanitarianly, or otherwise effective—whether they are neutral in those spheres, or if they are harmful. It will require also systems that determine related contributions to international relations, conflict resolution, security, world stability, and mutual understanding: a new exchange of ideas, and levels of communications and cooperation [9]. As the next essayist, Justine McGowan, demonstrates, many of our best efforts slip through M&E nets.

The doctor and the observatrice

We were snaking through the countryside between Tangiers and the town of Chefchaouen when traffic suddenly slowed. It was the first time in over an hour we had thought of much more than the car in front of us, in which a driver and a local fixer hired by our organization rode. I had been in northern

(Continued)

(Continued)

Morocco for six weeks as an international observer for the 2011 parliamentary elections, and my observation partner, Oscar, was a medical doctor from Guatemala. The previous day—Election Day—had been a long one, starting before sunrise, and we had visited as many polling stations as we could in the rural areas around Chefchaouen. Now we were on our way back to Rabat to regroup with the rest of the delegation, who had been scattered throughout the country, and write up our observations.

Oscar was a seasoned election observer, and as we drove in the light rain he was telling me about his last assignment in Haiti, where he would see cholera patients at night after spending the day meeting with political parties and activists. That combination of individual patient contact and care and being involved in a country's political process kept him from being too burnt out or frustrated by any one method of engagement.

The political climate was tense throughout North Africa and the Middle East, but in Morocco we had been able to attend many protests as observers without encountering the violence that had engulfed much of the region. But, after a couple of minutes in this random, middle of nowhere, traffic, we suspected that something had gone wrong. We started hearing shouting, but through the rain still could not see anything beyond the line of cars ahead.

Eventually, our car crawled forward and we could see there had been a serious accident. A passenger bus had lost control, skidded off the slippery road, and was now upside down, perched precariously above a trench around two meters deep. Injured passengers were strewn across the hill, mostly crawling, some moaning, and others shouting as they tried to pull out passengers who were still stuck inside the bus.

We took the scene in quickly and within seconds I was on the phone with the driver ahead of us to pull over and get out the car's first aid kit. Oscar and I threw off our election observation vests and hurried outside in the rain to start assessing the injuries. For the next blur of an hour, we went from one injured passenger to the next, with me interpreting in Arabic and preparing bandages and supplies and Oscar examining and triaging injuries. Very much in shock, the passengers were also quite confused as to from where these two foreigners suddenly appeared, shouting to each other in English, one speaking to them in the Syrian dialect of Arabic, the other consoling them in Spanish and French and treating their wounds.

A group of men next to us used a pocketknife and our flimsy scissors to cut the curtains out of the bus and build a make-shift stretcher to transport an old man who appeared to have broken his pelvis over the trench. Many of the passengers had very deep wounds and broken limbs, and we stayed treating patients until the ambulance finally came over an hour later.

(Continued)

(Continued)

As we finally climbed up the ditch to make our way back to our car, I noticed an elderly man squatting on the side of the road, staring peacefully in front of him. *"Keefak, amu, inta tamaam?"* I asked him, *"How are you, uncle, are you okay?"* He turned slowly and shot a toothless smile at me, some blood in his mouth as he had lost his last few remaining teeth in the accident. *"I'm fine dear, bless you both!"* he answered.

On the drive back to Rabat, we encountered a few cars that had passed the accident site and recognized us. They rolled down the windows and cheered enthusiastically, sometimes yelling their own blessings and appreciation. This was far different from the skeptical glances and reluctant discussions on the political process we had with community members over the last weeks. I had spent weeks in Morocco focused on supporting the objective of fair elections, and had made what felt like the greatest impact of my stay in less than two hours, on the side of the road, covered in mud and a bit of blood (Fig. 5.8).

Figure 5.8 Justine McGowan and colleague in the field. *Picture courtesy Justine McGowan.*

Justine McGowan

5.4 Measuring the immeasurable

Such systems will need new tools, and the evolution of existing tools. Tools that use formulae, models, or scores with sets of criteria to figure out if other good, or not so good, things are happening. Models that go beyond the numbers—that drill deeper and consider quality, as much as quantity: the universal, as much as the detail. Such new formulas come only from time in the field: this is the only way that the old systems can be adapted, evolved and modified into new systems which show that global health can deliver downstream effects, as well.

So, we need to ask new questions of performance measures: to ask if it would be possible or feasible to conduct, without too much extra effort, assessments of transcendent effects? And, if so, we need to ask if these results would jive with traditional values—with narrow measures: do the results fit with effectiveness, utilization, and uptake? Do they correspond with the number of people tested for infections, or the number of children taking antimalarials? What if the health is good, but the magic is not there—or vice versa; what happens if numbers say to do one thing, and instinct says to do something else? What should the threshold, the balance, be between the two? What if the health is dependent on the collateral; dependent on its capacity to attract funding? New questions such as these are only answerable through holistic assessments, that will reveal to both the internationals and to the locals the full picture.

Some effects are too nebulous for numbers. Sometimes, only the intuitive, a sixth sense, will tell you if they are smart ones. Or, perhaps not: could there be a way to identify in numbers, the threats and the strengths, the good and bad of infectious disease control on other levels—on every level? It is perhaps worth searching for a third way: somewhere in the no-man's-land between numbers, instincts, and *realpolitik*. To search for an alternative, an innovation—something that has not been done before. It is worthwhile breaking out models and breaking down systems: decoding and deconstructing the multilevel value of the best programs. It is worthwhile, perhaps, trying to evaluate dominos of respect: is the epidemic control or public health emergency program responsive to health needs — to nonhealth needs? Does it consider local values in relation to matters of culture, religion, society? Does it —and can it—adapt to them? Is the program locally-led—do locals have any say in decisions or in design—or was it externally imposed, without consultation? How does it affect the economy, the environment—and which is more important?

Ownership, adaptability, and all the sub-themes—the niggling ripples, undercurrents, afterthoughts, and glimpses—can all evolve into something real. Each, being channeled and captured; each morphing in to a question, or set of questions; each question with its own answers, its own scales. Scales that say, for example, the intervention is highly advantageous, diplomatically. Or another scale, that says: it is neutral, or it is a threat; is it benign, or malign. Scales of answers then emerge as well embracing and combining comments, explanations, observations, instincts, numbers—and everything in between.

The irony is seeing that the whole idea, from the start, is that global health is not about the numbers. That the elements of the diplomatic in epidemic or pandemic responses are beyond numbers. It is thus ironic to come back to the quantitative: the revolution turning on itself, as all revolutions do. But barefoot diplomats are needed to temper the quantitative with the qualitative: to weigh and judge in a way that that is not a slave to numbers. In ways that instead responds, respects, and values the nebulous—the ephemeral. To find a way for the quantitative to evolve in to the universal; to move from the utilitarian to the enlightened. To find a system through which the numbers can dovetail with the nonnumerical to produce holistic scores: magic numbers. The numbers that reflect the good: that the program is adaptable, effective. And the numbers that reflect the bad: that it is unsustainable, or insensitive (Fig. 5.9).

I searched for a way to get to that number, around the world. In San Francisco, there were programs about health education for HIV/AIDS—about teaching how not to get sick, or infect others. But these education programs, at the same moment, also advanced the social standing of the sick and educated society on the bad karma of stigma and of excommunication. Such responses—first in San Francisco, then around the world—at the time were becoming the beau ideal of global health programs. Health education was becoming a catalyst for social change, diversity, and tolerance as well: such programs were blindsiding malign intolerance; health education also advanced progressive thinking: on human rights, humanitarianism, fairness, and inclusiveness.

In San Francisco, epidemic control health programs therefore did more than meet the eye—more than they could ever be given credit for; more than the numbers could measure. But later, in Uganda, I saw the limits: the double-edged sword. I saw such programs as doing too much, too soon: saw benign liberal international values that were too progressive. Saw internationals pushing—rushing—for dramatic social reform; saw them challenging, too abruptly, centuries-old values, standards, and attitudes. Saw programs that were smart but lacking in cool; and saw draconian reactions: the exceptions that proves the rule. Saw, as schoolteachers are fond of saying, that it is possible to be too smart. Saw that one had to be careful, and cautious: careful, when trying to

Figure 5.9 All thanks to a monitoring and evaluation visit—new connections between locals and internationals who otherwise might never have met in remote South Sudan. *Picture courtesy Sebastian Kevany.*

get programs to do as much, on as many levels, as possible. And cautious, with sensitive subjects—with deep waters. Epidemic control efforts play, inevitably, inexorably, on the fringes of bigger pictures.

And so the search goes on for enlightened, renaissance global programs. For programs that bring communities together: that elevate the locals and the internationals and their organizations; that built collaborations and partnerships. The programs that are tuned in to local tastes and styles and preferences; that are set up in ways locals like. Programs that consider, for the first time, the ripples they produce.

From San Francisco to Uganda, and from Uganda back to South Africa: to HIV/AIDS programs in the bush, on the veldt. There, I wanted to capture the x-factor: to turn dreams into reality. To describe what was right, what was wrong—what could so easily be tweaked to advance, not just for epidemic control, but everything else as well. To troubleshoot, so that in the 21st century—if we are smart and cool about it—we can guarantee programs that are sustainable, culturally sensitive, adaptable, visible, and cost effective—that are holistic, as well as humanitarian—that tick all the boxes.

Figure 5.10 With modern tools and techniques, monitoring and evaluation can take place anywhere—even in clinics deep in the inaccessible Papua New Guinean jungle. *Picture courtesy Sebastian Kevany.*

Renaissance programs—programs that ensure the internationals are happy and the locals get what they want. Programs that win the hearts and the minds of policymakers, politicians and more importantly, the general population: programs that satisfy the aid skeptics, the health economists, the epidemiologists; that satisfy the man or woman in the street by letting them see the connection: the benefit to him or herself, as much as to the world. Programs that everyone across the board—from the right to the left, from the rich to the poor, across every social, economic, cultural, political spectrum—can see value in (Fig. 5.10).

5.5 Key messages

- Monitoring and evaluation has become a key theme of many epidemic control efforts, despite its reporting burden on local staff.
- However, efforts to capture results both help with knowledge transfer and surveillance as well as strengthening funding rationales.
- There is also an important anticorruption element to the reporting and reviewing of figures from the field.
- However, these figures may disguise the success, on more holistic levels, if interventions that do not always reach their targets.
- To this end, there is the need for the deployment of tools that try to capture (Fig. 5.11).

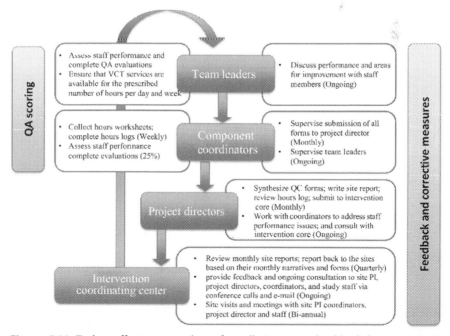

Figure 5.11 Today, efforts are made to formally integrate health diplomacy principles with monitoring and evaluation techniques.

References

[1] Rushton S, McInnes C. The UK, health and peace-building: The mysterious disappearance of health as a bridge for peace. Med Confl Survival 2006;22(2):94—109. Available from: https://doi.org/10.1080/13623690600620940.

[2] Mulley S. Donor aid: new frontiers in transparency and accountability. United Kingdom Transparency and Accountability Initiative; 2010. Available from http://www.transparency-initiative.org/wp-content/uploads/2011/05/donor_aid_final1.pdf.

[3] Lyman NP, Wittels, SB. No good deed goes unpunished: the unintended consequences of Washington's HIV/AIDS programs. Foreign Affairs; 2010. Available from: http://www.foreignaffairs.com/articles/66464/princeton-n-lyman-and-stephen-bwittels/no-good-deed-goes-unpunished.

[4] Martuzzi M, Tickner J. The precautionary principle: Protecting public health, the environment, and the future of our children. Geneva: World Health Organization; 2004. Available from: http://www.euro.who.int/__data/assets/pdf_file/0003/91173/E83079.pdf.

[5] Rubbini M, Tjivikua T. Health challenges and perspectives for sub-Saharan Africa. Int J Public Health 2018;63(9):1015—1016. Available from: https://doi.org/10.1007/s00038-018-1130-6.

[6] Jan S. A holistic approach to the economic evaluation of health programs using institutionalist methodology. Soc Sci Med 1998; 47: 1565—1572. Retrieved from http://www.sciencedirect.com/science/article/pii/S0277953698002287.

[7] Feldbaum H, Michaud J. Health diplomacy and the enduring relevance of foreign policy interests. PLoS Med 2010; 7(4). Retrieved from http://www.plosmedicine.org/article/info%/3Adoi%2F10.1371%2Fjournal.pmed.1000226.

[8] Fidler D. Reflections on the revolution in health and foreign policy. Bull World Health Organ 2007; 85(3). Available from: http://www.who.int/bulletin/volumes/85/3/07-041087/en/. https://doi.org/10.2471/BLT.07.041087.

[9] Kevany S. Global Health Diplomacy, 'Smart Power', and the New World Order. Glob Public Health 2014. Available from: https://doi.org/10.1080/17441692.2014.921219.

CHAPTER SIX

War and peace: barefoot diplomacy as military adjunct

Abstract

The increasing cost and limited demonstrable effectiveness of many early 20th century conflicts, combined with increasing public awareness about the explicit and implicit costs of national and international conflict, have led to calls for alternatives to hard power initiatives. Alternative systems of strategic or tactical pursuits, combined with the protection of spheres of influences and hegemonies (by both world powers and other countries), now include the strategic deployment of health and infectious disease programs to win hearts and minds or gain strategic allies at the national or subnational levels. In this chapter, we explore both the risks of such strategies, including the subjugation of health programs to strategic aims, and the benefits such as a compelling rationale for the diversion of defense in to development.

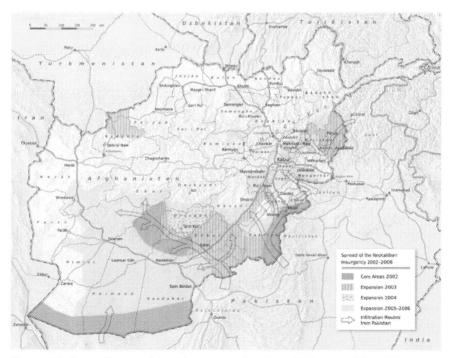

Figure 6.1 Can military and health presence be aligned? A map of the riskiest provinces during the Afghanistan conflict.

Barefoot Global Health Diplomacy.
DOI: https://doi.org/10.1016/B978-0-12-818681-7.00001-7

6.1 Unlikely bedfellows?

The centerpiece of barefoot diplomacy and perhaps its true raison d'être is the mitigation of conflict-related suffering (Fig. 6.1). An end to such suffering could be the greatest potential downstream contribution of health to humanity, by advancing the perennial search for less violent ways to resolve national and international conflict. The search for a strategic function of armed forces in humanitarianism, and vice versa, can be complemented by mutual strategic contributions from epidemic control efforts—global health personnel, and militaries, working together by combating extremism and terrorism through HIV/ AIDS or tuberculosis or malaria programs—enhancing global cooperation and security through tuberculosis treatment or malaria prevention.

The search is on for alternative or for new ways of resolving international differences; of challenging medical paradigms in the international relations context due to the disappointments, tragedies, and failures of old models of power and influence. We have, perhaps, been using ineffective systems, in both realms—with outdated solutions, so the same problems persist. The world is thus perhaps looking for new systems of stability, seeking order amidst chaos. But systems are needed that enhance and embrace infectious disease control's contributions—evolving its role in to an alternative or complement to military responses. Smart approaches—without compromising identity, standards, vision or security—can, in the 21st century, be just as effective as the military, via partnership in conflict resolution and peacekeeping.

We also need new ways to pursue stability because the world is intrinsically chaotic and confrontational. Wars and conflicts are as much a part of humanity as ill health and disease; their persistence and constancy are juxtaposed with the often-transient effectiveness of traditional solutions and responses. Hard power has evolved over the centuries; there has been an ever-increasing capacity for its deployment, but war tech often becomes dated as soon as it has been created. Yet even with, or perhaps because of, this ever-increasing capacity, there has been declining approval toward warfare; throughout history, attitudes toward violence and conflict have been evolving, as well. Tolerance for extreme forms of conflict and weaponry is changing; national and international consensus drawing invisible, yet increasingly narrow, lines of acceptability.

Just as the tools of war and conflict have been evolving, so attitudes have been moving toward civility—yet, war itself seems to have become less orderly and predictable, more violent, and less explicit or structured. Wars are no longer fought between professional soldiers and standing armies; instead, they are focused and often perpetrated by nonmilitary civilian entities. Military prestige has been affected; nimble, civilian, and global guerrilla wars are often beyond the response capacity of conventional militaries.

Global society now questions the value of military responses because of dawning realizations that, on occasion, such investments can actually cause the conflicts they are designed to resolve and prevent—particularly when misused by morally bankrupt governments, and particularly in the developing world. Improved global communications, visibility, and awareness also have a part to play; since war documentation has improved via the journalists and videographers of Vietnam, there has been an evolving awareness of the grim realities of conflict. The general public can see in front of its eyes how traumatic and brutal war actually is; modern war is thus often stripped of romance and adventure, laid bare of Hollywood and cartoon glamour (Fig. 6.2).

The decline in perceptions of the value and effectiveness of war is also consequential on the development of devastating technologies; technologies often too devastating to use, such as the era of mutually assured destruction ushered in by nuclear weapons. The 20th century was the first

Figure 6.2 In Iraq, site visits to infectious disease clinics are security-intensive affairs, requiring armored cars, confidential agendas, and a large dose of diplomacy. *Picture courtesy Sebastian Kevany.*

time in history that advances in technological and scientific capacity for arms paradoxically then made their use impossible. Nuclear weapons instead created global stalemates: made military might too often irrelevant and ineffectual to the strategic paradigm.

Threats of nuclear war today seem dated and passé, and yet they still control a power over the collective consciousness: a threat of a constant, looming presence; a shadow of doom. The shadow is, perhaps, also a fear of conflict and its destructive, all-consuming power. Could this fear be a deep, increasingly widespread catalyst for efforts to find new ways resolve and vanquish threats to civilization, to peace, and to human rights? Perhaps, we can try and resolve these threats by lateral thinking, by new ideas, and by breaking malign economic loops of investments. That same money, time, and energy could instead be spent, on occasion, via militaries assisting with the health of the world's poorest people—in ways that also provide strategic and security returns.

In this context, the conflicts in Afghanistan and Iraq were occasionally resonant of the Vietnam War in their inconclusiveness. They also highlighted the monetary expense the wars brought. The 21st century thus dawned with a violent reminder that those who ignore the mistakes of history are destined to repeat them; in the age of technology and information and global awareness, the global zeitgeist is less tolerant of the costs of hard power.

Yet public distress and disenchantment are reaching such a crescendo that governments are now, maybe mistakenly, reluctant to set foot in possible repeat scenarios. Because despite the reluctance, the disillusionment, and the pacifism, the threats to national and world security persist. The threats are not going away just because dated responses have become unfashionable or ineffectual; the threats are a constant in an ever-changing, ever-unstable world.

In Afghanistan, and also on a couple of missions to Iraq, I saw the old, unresolved threats [1]. Saw old approaches struggling, and I saw infectious disease control efforts rising to the occasion. Saw global health do as much for peacekeeping and international relations and conflict resolution than militaries: malaria programs that were adaptable, enlightened, locally owned. Infectious disease control programs modified to function in extreme, insecure regions: in remote mountain villages, where no other international presence was welcome. But the epidemiologists were welcome, because they had tuned in to local culture and society; accepted and embraced through sensitivity and awareness. I thus saw programs that changed perceptions of internationals but that transcended identification with any one country, or set of beliefs.

Programs that unified and included also presented alternatives to visions conjured by remote, isolated extremism. In Afghanistan, and in the cold mountains of Kabul, there were inadvertent, strategic effects of infectious disease control. Hidden, accidental, forgotten effects beyond health gains; the counterbalance. This despite the similarly accidental, but well-exposed and publicized, costs of the aid paradigm: the demoralizing, inexorable corruption. The arbitrary ineffectiveness and shady politics, the back-room deals, and wasted effort: nullified by the elevation of the importance—the value, significance and effectiveness—of investments in health. In Cambodia, Dr. Michael Baker observed, in similar fashion, that broad benefits of medical aid may trump narrow costs.

The sword, the scalpel. . .and the toothbrush

In 1997, I was the Senior Medical Officer for a team from Naval Special Warfare Unit One and Special Boat Unit 11 that was tasked with providing village level health care in rural Cambodia. The Medical Civic Assistance Program (MEDCAP) was part of a joint training mission the US Navy held with the Royal Cambodian Navy, the first US Navy official visit in over 22 years.[1] This took place after nearly 20 years of relative darkness and violence, which had crushed Cambodia and its people. The Khmer Rouge regime had killed many people, dismantled services, destroyed infrastructure, eliminated teachers and educational opportunity, and driven the populace out of the cities. The country we visited was literally on its knees in terms of being able to provide for its people, let alone conduct business or health care. The infrastructure was devastated, and there was an internal struggle in the government by factions of the new United Nations peacekeeping effort, which imposed joint governance, hampering services to the people.

Our US Navy medical team worked with local medical providers, gave vaccinations, conducted basic lab tests, treated injuries and illnesses, provided medications, and performed simple surgical and dental procedures. There were also physical therapists, medical educators, and public health trainers in the group. They worked with village people to promote rehabilitation from injury, prevent illness from contaminated water, and to educate about safe use of pesticides. There were over 3000 attendees at our 2-week clinics.

(Continued)

[1] The Medical Department in Military Operations Other Than War. Part II: Medical Civic Assistance Program in Southeast Asia https://academic.oup.com/milmed/article-pdf/164/9/619/.../milmed-164-9-619.pdf

(Continued)

The dental detachment was very busy with pulling teeth and dealing with dental emergencies in people who had no oral hygiene or health care for many years. But the providers—especially the enlisted navy corpsmen (medics)—made a tremendous effort to make two other contributions: they provided simple dental equipment and training for local providers, and they gave out toothbrushes and taught oral hygiene to the kids in each village.

I am not certain about (in fact I am somewhat skeptical of) the real return on investment (ROI) in short term military humanitarian missions such as MEDCAPs. With great effort, we think we facilitated some local health improvement which could be sustained, but I do not feel this occurs often with short term deployments. The real superior ROI in humanitarian efforts is when the US military is the 911 force which provides quick disaster response: medical support, search and rescue, damage assessment, shelter, security, and life-sustaining supplies—as was done after the tsunami that hit Banda Aceh, Indonesia in 2004. US military personnel were the first responders, and US image around the world soared.

So why do I provide this story—where I worry that our short term MEDCAP mission had provided perhaps little value to the people, and not much ROI to the US? Because of the surprise I experienced when other military civil affairs groups reported on their visits to areas they had deployed to previously. They returned to the same area some time later, and were greeted by a slightly older group of children showing broad smiles, waving their toothbrushes (now quite worn down), and greeting US military team personnel like favorite relatives.

I offer this vignette as I observe the wreckage the United States leadership and its military have wrought in the Middle East during 18 years of war, to teach the reader about "opportunity cost." I learned about opportunity cost, "the result of making a particular choice compared to the return from the most valuable choice not taken," because the military sent me to some business school type lectures and courses. So, how could we optimize that "opportunity cost" to provide a better return of investment?

We have many opportunities in which I imagine a different scenario, and a better opportunity cost. For example, where for the price of one cruise missile hitting a Syrian runway (and not changing much of anything)—a lot more children around the world would be smiling, waving their worn down toothbrushes, and thanking the people and military of the US for the efforts on their behalf. There we would find a better return on our investment, as we did with our MEDCAP in Cambodia (Fig. 6.3).

(Continued)

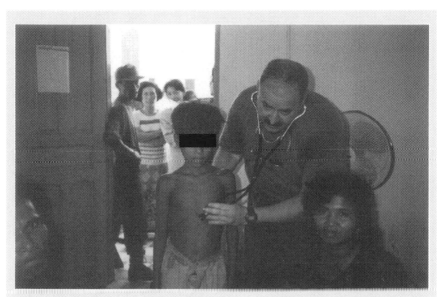

Figure 6.3 Military operations other than war (MOOTW) in Cambodia. *Courtesy Michael Baker.*

Michael S. Baker

6.2 A new stage for security and stability

Changes in attitudes to war—the changing standards of style, levels of tolerance for conflict—have also been catalyzed by global connectedness, through the transparency of links between conflict, disease, and poverty via education and enlightenment. There has been a change through realizations that global inequalities are the causes, as well as the effects, of war and destruction; here are realizations that by addressing ill health and disease, the levels of global hatred, jealousy and resentment also decline. Realizations that declining adverserialism can be achieved not only through the direct efforts of epidemic responses to redress global inequality, or by the essential use of military force, but also by developing new roles for militaries as agents of humanitarianism.

Such new roles require a new nexus. They can be achieved through National Armies for Global health [2]; through innovative uses of militaries to fight diseases, as much as wars. The fight of viral and conta-gious enemies today represent just as much a threat to world stability as extremism and terrorism; a new nexus is needed for fighting battles and wars that addresses extremism, terrorism, and contagion all at once. Such innovative uses of military might act as an inspiration for the 21st century: may act as a step toward a world in which applying the military against health threats also addresses other strategic or stability concerns.

Utilizing armies for fighting disease is also a response to terrorism—a way of avoiding the "Iraqs and Vietnams." A way of avoiding those mili-tary engagements that may, ultimately, do more harm than good from security perspectives; it is a new departure that leverages and taps in to defense budgets, which are significantly higher than those for infectious disease control. Yet the needs and demands of the latter remain constantly beyond supply; today, the supply and demand of the military and the altruistic are gradually equilibrating.

Equilibrating through barefoot diplomacy, through dovetailing global health and strategy; through policies and practices, world events and govern-ment departments, which are no longer operating in isolation of each other. Au contraire, the concept of stove-piping purviews such as defense, develop-ment, and health is becoming increasingly inconceivable—even if epidemic control or public health emergency efforts and the military are not always sit-ting at the same table—even if they do not always see eye to eye.

More and more, global health and international militaries are on the same page through improved communications; through ever-increasing opportu-nities for overlap and collaboration; and through mutual through exposure. Exposure, as individuals and departments, soldiers and epidemiologists, become increasingly mutually aware—increasingly omniscient. Soldiers, beginning to see health threats as opportunities for their militarized containment, just as infectious disease control workers see via their peripheral vision threats of con-flict, and the opportunities for their resolution. Both sides are thus seeing opportunities and synergies that contribute beyond their specific fields of endeavor; that can consider and fix broken parts of the world machine.

Global health is thus contributing to the pursuit of goals, which was once the exclusive realm of the military—however, it is pursuing cautiously. The interplay of the medical and the strategic is not uniformly benign. Inevitably, it is a double-edged sword—on the one hand, it risks the security of epidemic control programs and personnel, local, and international. These are new demands placed on doctors and nurses and community health-care workers. Different demands—but they require the same elements of character that first

Figure 6.4 In unstable South Sudan, epidemic control motorcades often partnered with, and took on the appearance of, more martial expeditions. *Picture courtesy Sebastian Kevany.*

brought those same nurses, doctors, and health-care workers to the field. Such demands also require that new collaborations be governed by principles, and that the limits to infectious disease control's effectiveness in nonhealth realms be recognized. The eternal need for militaries and armed forces still has to be accepted, even by the aid brigade [3] (Fig. 6.4).

Global health and militaries are, then, both evolving in to new roles: Military change has been catalyzed by a paradigm shift away from their 20th-century roles of soldiers and standing armies. There is now a shift, for example, toward military drone technology—leaving manned styles of intervention abruptly redundant. Robots are now often eclipsing the old human systems of warfare—including the soldiers, sailors, and airmen. The latter, though, are adapting to survive: adapting in benign, humanitarian ways; to adjust to roles of diplomacy and rescue. Such roles are providing them with their most rewarding tasks—increasing self-esteem, and decreasing personal mental and physical, trauma [4]. Infectious disease control roles for the military are thus a smart, adaptable, Darwinian evolution: from Ebola outbreaks in West Africa, to national and international health security, an evolution is occurring to leverage the responsiveness and reach of military and security apparatuses, unique in their worldwide scale and scope, to protect health.

Military contributions to health protection are complemented by infectious disease control's contribution to human protection and world stability. Barefoot diplomacy is contributing to that, as well: military roles in global health are playing on the edges of soft power, as well as of humanitarianism, by attracting a diverse and pluralistic world to better opportunities and a sense of dignity [5]. Smart approaches, without compromising medical integrity or independence, are likewise contributing to broader values and aspirations through inoculating and vaccinating against extremism [6].

Contributing not only to national defense and security but also to global stability: contributing at global as well as national levels. But perhaps barefoot diplomacy is thus beyond mere bilateral soft power, is more than just cooperation in lieu of coercion: perhaps it is capable of contributing beyond nationalistic efforts at attraction or persuasion to a diplomatic tool kit that achieves the goods of global or national communities using every means possible [7]. A tool kit that includes epidemic and public health emergency responses—that addresses the causes as well as the symptoms of global animosity; that understands that deprivation and inequity lead to conflict; that understands that poverty and disease can lead to desperation and risk.

Global health is thus contributing to the strategic priorities and concerns, not only of donor countries but also of the global community; it is pursuing common values and aspirations that all countries and populations—locals and internationals—agree on. It is thus contributing to transcendent principles, global goals: to greater quotients of world security and conflict resolution; to the eradication of global terrorism and violent extremism. It can contribute to fewer dictatorships, genocides, and epidemics; to less environmental degradation; to greater communications, or better balances of power. It is also contributing to improved international relations and global economic prosperity; and, often, to hard fought, precarious independence.

But substitutes for, or complements to, hard power—such as infectious disease control, even with its nuanced, inspired overlaps—still cannot do it all. Still cannot resolve all the international tensions that cannot be taken out on football pitches: world conflict is inevitable, often are a part of human DNA and survival. In some cases, wars may perhaps be necessary—and whereas they are unlikely ever to be benign, they can be fair and just. There is thus legitimate investment in defense, in "good" wars—but there is the need for balance between the use of hard and soft and smart power. There is thus also a necessity for balance between efforts, investments, and collaboration across the development, diplomacy, and defense spectrum: a need for alignment, harmonization, and synergy optimization with each mechanism supporting and advancing the others. As Nurse Coral Andrews explains, defense forces can be powerful forces in public health emergencies as much as in the field of battle.

Duty calls

The path to a career in Global Health Diplomacy (GHD) is not always intentional or self-directed. Circumstances emerge that require a depth and breadth of skills, tested and untested, to navigate and deliver value during large-scale, international events. This is what launched my path.

On December 26, 2004, a tsunami struck the Aceh province of Indonesia following a 9.1 magnitude earthquake—225,000 people died. In addition to extensive loss of life, this Southeast Asia tsunami produced multi-national devastation. In addition to Indonesia's impact, the world soon learned that other affected countries included Thailand and the Maldives.

In January 2005, I received a call from the US Pacific Command requesting immediate executive staffing assistance in the Command Surgeon's Office as the acute phase of a large-scale, international humanitarian assistance effort got underway command-wide. To step back, my career began as a Nurse Corps Officer in the US Navy in the late 1980s. While I had transitioned to the private sector following active duty, I remained in the US Navy Reserves: as a senior officer, I had worked in support of multilateral engagements in other Directorates at the US Pacific Command and could "hit-the-ground-running." The 24/7 response required bandwidth from a team that could capably synthesize complex variables, work in cooperation with multi-national teams, and have the discernment to work within the governance framework of international relationships and partnerships.

The US Pacific Command, where I was assigned, was renamed the United States Indo-Pacific Command in 2018. To set the stage for the variables that needed to be synthesized in the diplomatic response plan to the affected region, it can be beneficial to define the overlapping domains and practice disciplines as described in the Command's area of responsibility.[2]

1. There are few regions as culturally, socially, economically, and geopolitically diverse as the Asia-Pacific.
2. The 36 nations comprising the Asia-Pacific region are home to more than 50% of the world's population, 3000 different languages, several of the

(Continued)

---

[2] From http://www.pacom.mil/About-USINDOPACOM/USPACOM-Area-of-Responsibility/

(Continued)

world's largest militaries, and five nations allied with the US through mutual defense treaties.

3. Two of the three largest economies are located in the Asia-Pacific, along with ten of the fourteen smallest.

4. The area of responsibility includes the most populous nation in the world, the largest democracy, and the largest Muslim-majority nation.

5. More than one third of Asia-Pacific nations are smaller, island nations, including the smallest republic in the world and the smallest nation in Asia.

6. The region is a vital driver of the global economy and includes the world's busiest international sea lanes and nine of the ten largest ports.

7. The Asia-Pacific is also a heavily militarized region, with seven of the world's ten largest standing militaries and five of the world's declared nuclear nations. Given these conditions, the strategic complexity facing the region is unique.

Global health diplomacy (GHD) is a dynamic practice; complexity and dimensions grow as the international arena evolves. The numbers of state and nonstate actors (those in-country, and others who joined in-country over time) expanded the disaster and humanitarian response considerations in the areas of diplomacy, foreign policy, international security, health security, culture, and strategic communication. Below, I have highlighted a few reflections. Each reflection is paired with a key component of the practice of GHD.

Diplomacy: The affected country takes the lead in deciding the amount and types of assistance they want. Work within the governance framework, while leveraging Embassy and Attaché' personnel.

International security: The safety of response personnel must be balanced with the need to provide immediate assistance. There are publicly reported sources describing the presence of militants in Indonesia that could have posed a risk to state and nonstate actors on-the-ground.

Foreign policy: When teams of countries come together in an acute situation, it is still imperative that "good hearted intentions," state or nonstate, operate within a legal framework to abide by international laws and treaties.

Health security: Leverage disease surveillance resources, such as the Centres for Disease Control (CDC), for knock-on considerations. In a tropical climate, the sitting water presented concerns about secondary devastation from diseases. Access to drinking/potable water resources had been disrupted.

Cultural considerations: The massive number of deaths presented challenges in complying with cultural norms for timely burials.

Strategic communications: The growing presence of nonstate actors in international relations necessitates that a uniform communication platform be

(Continued)

(Continued)

available in large-scale international disasters. Technology has likely helped to fill this gap.

Indonesia once again faced a devastating tsunami on September 28, 2018. Yet much has changed in the world since 2004. We continue to experience changes due to globalization that are shifting the nature of relationships across the globe, and bringing new entrants into the international relations policy arena: security threats resulting from terrorism necessitate considerations of safety when engaging both diplomatically and on humanitarian levels. Innovation in health is also expanding the reach of global health initiatives: technology has the potential to change the way that GHD is practiced in the future. In that context, I wonder how the response to the 2004 Southeast Asia tsunami be similar, or different, if I was called to duty today?

Coral T. Andrews

6.3 Mutual sacrifice? Lesser evils, greater goods

Optimal synergies, in this regard, are governed by formulae and principles: the same principles, in a different manifestation, that represent the essence of barefoot diplomacy. The synergies that elevate infectious disease control to an effective alternative to hard power: smart power, acting as an alternative to military coercion, and used to influence states' or political bodies' behavior and interests [8]. Principles, also, that govern better capacity within militaries for global health; that embrace training and connections between epidemic control and intelligence and influence. Therein lie the principles that connect epidemic response efforts with other global forces for good; that align fundamentally civilian conceits, such as barefoot diplomacy, with military systems of medical engagement. These are the principles that consider inter-organizational alignment and communications—that are mindful of both the geo-strategic, and the geo-political [9].

Principles, then, to ensure that barefoot diplomacy wins hearts and minds: the hearts and minds, in particular, of the skeptics. Such ad hoc diplomacy can thus be a unifying force: postpartisan, uniting hawks and doves, right, and left. It can bridge the growing divisiveness and divisions that threaten the integrity of modern democracies and global balances of power: smart approaches can provide for the skeptics the security returns, strategic returns, and many other returns on investments.

But the recognition and use of global health's wider effectiveness and potentialities brings with it new risks: that future conflicts will be cast as armed social work, more than traditional battlefield confrontation [10]. Due to infectious disease control's inexorable links with conflict and its health effects, there are new risks to be avoided—epidemic response efforts do more than meets the eye: they are often an international presence in conflict zones, which results in creating oases of stability and refuge. An epidemic response presence often thus means that there is sometimes one less opportunity for rogue organizations to commit atrocities and war crimes, as any kind of global presence can serve as another pair of witnessing eyes and ears.

The presence of global health efforts within military purviews also holds the possibility of armies and navies becoming institutions whose intelligent, multilevel pursuit of human protection demotes violent conflict to a last resort. In West Africa, international armies have pursued international defense and security not through ballistics, but by responding to Ebola outbreaks—by building local clinics, saving local lives. In doing so, they are easing local and international pressures by benignly occupying military attention and energy, by preventing associated restlessness, and by improving the image and morale of participating militaries. By being, essentially, smart: by recognizing the continuous, inevitable blurring of lines between responsibilities and organizations, departments and perspectives, professions and purviews.

Breaking down walls and aligning priorities is difficult to achieve, as it requires expertise. In conflict zones, should epidemic control, for example, focus on places and populations in which it can affect optimal strategic, conflict resolution, or peacekeeping impact? Would that result in attracting more funding and support—or should efforts instead focus on less widely supported, less compelling, pure altruism; on the places and populations that need epidemic response assistance the most? Global health's use and effectiveness in conflict resolution will thus require changes in resource allocation decisions—will require new choices on which programs, places and populations to invest in. Difficult choices may need to be made, unless the places and populations overlap, and those of greatest strategic importance are also those affected most by conflict and report the worst health outcomes.

Global health's dual role will require sacrifices on both sides, therefore, but a vision that provides alternatives to war will also attract new support. Barefoot diplomacy will, in essence, save more lives than if choices are made purely on the basis of health economics and utilitarianism: cost-

Figure 6.5 In Northern Sudan, tanks and ballistics were a constant backdrop, with wide range of collaborators required to make global health efforts feasible. *Picture courtesy Sebastian Kevany.*

effectiveness can be trumped by arguments for support beyond the vision of health economists. Arguments based on militaries operating through infectious disease control, and vice versa, can achieve multi-level goods: the global good and moral good; strategic good and medical good. Both sides can contribute to the benign and altruistic, through creating new frames of reference for societies and individuals. Through this, both sides create new role models for the millennials and Generation Z: humanitarian warriors, as the next generation of Hollywood heroes: responding to the imagination and imperatives of 21st-century society (Fig. 6.5).

I have been lucky enough to see the new beaux ideals, the new heroes, around the world. In Iraq, I saw both humanitarian military corps, and the flip side: saw nurses and doctors, local and international—through their style and diplomacy, sensitivity and awareness, bravery and valor—doing more, inadvertently or intentionally, than helping the sick. Saw them optimizing the communication and collaboration ripples of their work, and the downstream effects of international presence and local acceptance. Saw them also pursue socially integrated, culturally adaptable, religiously sensitive programs that hinted, as well, at invisible international relations outcomes.

Similarly, in Egypt, I saw precarious instability—Tahrir Square, and coups d'état. But I also saw stability and continuity in epidemic control: the tuberculosis and HIV/AIDS programs in Cairo and Alexandria, in the lion's den. Epidemic response programs that persisted amidst the chaos as we drove through the packed, militant throngs in downtown Cairo: the programs which persisted as we walked through military cordons, around government buildings, to enter the Ministry of Health. These programs, inevitably, overlapped with wider determinants through their commitment to service delivery, their pursuit of safety and continuity—a counterpoise to danger and instability. They counterbalanced chaos through the fatalistic, unflappable Egyptian public health doctors and nurses; through those who phlegmatically ran their clinics as the cauldron of mob and military boiled outside.

In North Sudan, also, I saw the work of doctors and nurses and community health-care workers far beyond Khartoum: the isolated, remote altruists, tied to their locales in spite of their training and status. Doctors, nurses, and community health workers welcoming the international just as they helped to relieve international tensions: locals were working and functioning in an era of conflict and a milieu of isolation and inaccessibility, yet were a welcoming presence to those internationals, myself included, who otherwise would have been banned from entry [11]. In North Sudan, epidemic control thus brought internationals to off-limits regions: to places as hard for internationals to get in to, as they were for locals to get out of. Infectious disease control investments and efforts beyond Khartoum were thus justified—not only by altruism—but by their contributions to conflict resolution, access, diplomacy and international relations.

One other story springs to mind: old man stood under the sun in the middle of the road in the middle of a village in Zimbabwe. We both discussed global health, and life: he told me that as he grew older and looked back on his life, he had realized the hidden duality of his existence. That what he had been doing for the village, was actually also for the world—but that, when compared with what he thought he had been doing, turned out to be two different things. He now looked at his contributions to village life which were only clearly visible in retrospect that they went beyond functionality: he had realized that they were as much about watching for threats to the village viewed from mountain hilltops, as they were about herding his sheep.

The old man, perhaps, captured the essence of the implicit truths and explicit facts of so many other lives, duties, and actions. He captured the explicit outcomes and implicit effects of global health: later, in Papua New Guinea, standing on top of a hill, I saw hospital ships both providing

Figure 6.6 The old man at the end of the road. *Picture courtesy Sebastian Kevany.*

health and extending international relations olive branches. The ships were perhaps the modern alternatives to invasion and conquests previously in pursuit of natural resources; now the ships blurred the line between medical and political, the strategic and the economic. Dr. Sheena Eagan experienced these same tensions, at first hand (Fig. 6.6).

Armed humanitarians

As I looked out beyond the dirt of my makeshift window, I saw a long and disorganized line of people that tendrilled on endlessly underneath the hot sun. There were thousands of them—mothers clutching their babies against their chests, elderly men with crutches barely held together by tape, and young children running back and forth along the dusty road. It all gathered into a confusing mixture of cries of play, tears of pain, and resigned heavy silence.

Near the front of the line, I noticed a young woman with lines of worry on her face as she quietly and intently held a bundled baby in her arms. The blanket, which was bright blue and freshly laundered, stood out against the desert landscape of dust and sand. As I scanned over the crowd, my eyes always seemed to wander back to this woman and her bundle of blue. I knew this sight well—too well. This

(Continued)

(Continued)

mother's face had a look of worry that I had seen often throughout my career as a pediatrician: although there was youth in her eyes, her expression was heavy. But even so, there was a resolute determination in her stance; she was not here for herself, but for her child. Something was wrong. Something was very wrong.

By chaos or by destiny, she was escorted into the small room where I was seeing patients. Seeing that familiar look of deep concern, I attempted to greet her with the same compassion and confidence that I would when walking into my exam rooms back home. I am a doctor. I should be able to help, right? But this was not home, and she was not the patient that I am used to seeing. She looked at me quizzically, clearly unable to understand my words of comfort. She quickly presented me with her beloved bundle of blue. I was right; she was here for her baby. I looked down at a small face peering up at me, her wide eyes deadened by heavy, dropping lids. It was emaciated and quiet—unnaturally quiet. The baby was freshly washed and wrapped with love, but clearly, she was sick. I noticed the yellowish tinge upon her face almost immediately. Unable to communicate with the mother, I quietly began my examination.

Unwrapping the baby revealed the same pot belly that I had seen time and time again, a sad reminder that diarrhea and dehydration continued to claim young lives in these parts of the world. But it was when I placed my stethoscope on her small, gently heaving chest that I was confronted with what was really wrong—a loud roar, a congenital heart defect. If I was home, I would call for tests and order consultations with specialists; I would be able to help. I am a doctor. I should be able to help, right?

But I was not at home. There was almost nothing I could do. Although there was a referral system, I knew that this woman would not be able to travel the 300 kilometers to the nearest hospital for the care her child needed. I resigned myself to reality, like I had done so many times before. I treated the baby's diarrhea and found a translator to recommend follow-up care. Our one-day clinic had nothing to offer them. Yet when the day ended, and we packed up our supplies, many were still aware of the work that they had done.

The mission was a success to our command—we won hearts and minds. We achieved our strategic goal. But, what about the medical goal? Had we improved the health of this population? Could have done more for that mother and her child? Maybe. Maybe not. This was not the first time these questions plagued my mind. I knew it would not be the last either. At this point, my only solace was a resignation that was being pulled along by a thin shadow of hope.

The next morning, we moved to another village and then another, hopping across this impoverished land of dust and kindling. I saw more cases that

(Continued)

(Continued)

I could diagnose, but not treat. Still, that one child weighed heavily on my mind with her cracked lips, her listless limbs, and the roar of her heart. I became a doctor because I wanted to help people. It is the same reason that I decided to join the military. I wanted to help those who needed it and thought that military service would enable me to help people all over the world. I am a doctor. I should be able to help, right?

Sheena M. Eagan

6.4 Principles, policies, practices

Yet blurred lines, as we have seen, result in risky partnerships as they create cases for expansionist powers to encroach on sovereignty through health. They create risks also in terms of structuring the chains of command, and in terms of who is in charge. Blurred lines can thus result in risks in terms of prioritization of agendas; in terms of clashes of organizational cultures. Often, medical personnel can cross such lines by attempting to work within the formalities and hierarchies of armies and navies: unexpected role reversals, with nurses inadvertently peacekeeping in Iraq, or soldiers unexpectedly nursing in Sierra Leone.

Blurring lines between health and strategic realms is also risky in terms of security—due to the threats to personnel, and unwelcome associations putting them in the line of fire. There are also greater human security threats as a consequence of the augmented presence of medical or paramedical workers in conflict zones: the last two decades have been the riskiest in history for aid, development, and global health personnel [4]. Medical professionals are more likely now than ever before to be kidnaped or shot, and there are more instances of hospitals being caught in the crossfire. Local and international nurses, doctors and community health care workers have, too often, been caught in the wrong place at the wrong time; there are increasing demands on their professional duties, as a result of new occupational hazards.

Such new risks demand that the integration between infectious disease control and militaries is managed effectively, and that the overlap between the strategic and the altruistic is carefully calibrated. Managed integration between both is necessary to help epidemic control efforts resolve issues that armies sometimes cannot: managed integration, through the articulation and application of shared operating principles, can produce programs that operate in smarter, cooler ways. It can produce renaissance infectious disease control, public health emergency, and epidemic response efforts that leads us to the brink of an era where global needs, concerns, and imperatives can be dually-addressed.

A renaissance within global health and military efforts can thus lead us to an era where the vast apparatus of defense is partially reoriented to social justice—toward access to health services in poor villages, and in war torn lands. This can lead us toward health for both the excluded and the oppressed: toward an era which furthers efforts to limit, avoid or prevent violent conflict, while still achieving the world's strategic goals and imperatives, via HIV/AIDS, tuberculosis, malaria or other infectious disease programs: an era of smart health, as much as of smart bombs (Fig. 6.7).

Figure 6.7 In Papua New Guinea, military hospital ships form many nations competed for local attention amidst natural resource extraction overtures. *Picture courtesy Sebastian Kevany.*

This evolution could, also, lead us toward an era which is built both on operating principles and groundswell: on the zeitgeist of public opinion. Consigning past misadventures to history—those which the vox populi questioned—will demand new partnerships between military and industry, politics and infectious disease control. Partnerships are indeed evolving from all sides: armies which are now developing capacity for military operations other than war [4]; continuing to develop the capacities to promote peace, defuse aggression, and resolve conflict, in lateral ways. Another example lies from in industrial, switching focus toward investing in medicines, and from the political building momentum for new ideas on strategic ideals.

Global health is also becoming smart through multilevel organizational collaboration—through programs which have been selected and designed based on their capacity to be effective and appropriate in the strategic as well as through the epidemiological contexts (without any diminution of the latter). Programs that capture the essence of enlightenment through strategy combined with altruism; programs that are governed by principles that generate those effects, every time [12]. Principles that also govern altruism's moral obligation to respond to conflict; through principles that ensure infectious disease control agencies calibrate efforts to leverage their geostrategic and geopolitical access to off-limits regions and populations.

Principles, also, that ensure any augmented medical resources based on such collaborations are allocated across places, programs and populations which are based on joint, mutually reinforcing, strategic and medical goals: principles that help medical organizations and initiatives to develop an awareness of the strategic and security implications of their work. Principles also that avoid health interventions are likely to incite religious or social, economic or cultural, sensitivities—no matter how cost-effective they may be. And principles that ensure that above all, infectious disease control addresses the underlying causes of resentment, discontent, and extremism via addressing poverty, alienation, and disenfranchisement.

Global health is thus slowly starting to play its part. Often, it results in playing more than its part: foreign aid was once representative of, or defined by, the arming of poor countries. Foreign aid was also once synonymous with military industrial expenditure; but, nowadays, does foreign aid mean infectious disease control? If so, it means ideologies are advanced, and proxy wars fought, through medicines. As the warship sails away, the hospital ships sail in—quite possibly achieving the same ends (Fig. 6.8).

Figure 6.8 In wartime Iraq, infectious disease control programs, and their personnel, built many unexpected bridges between locals and internationals. *Picture courtesy Sebastian Kevany.*

6.5 Key messages

- Hard power is an essential tool in preserving world stability and security. However, it cannot be successful as a sole entity.
- This has led to the development of smart power techniques whereby defense and military forces engage with public health emergencies and other efforts.
- In turn, this can lead to greater regional stability, as well as the resolution of potentially life-threatening global epidemics.
- Military engagement in global health is also relevant in conflict resolution, where high related effectiveness has been evident in the past.
- Addressing health and epidemic issues may also help to disrupt the root causes of extremism related to international inequity and resentment.
- However, care must be taken to ensure that military health efforts do not inadvertently aid extremist or otherwise hostile populations (Fig. 6.9).

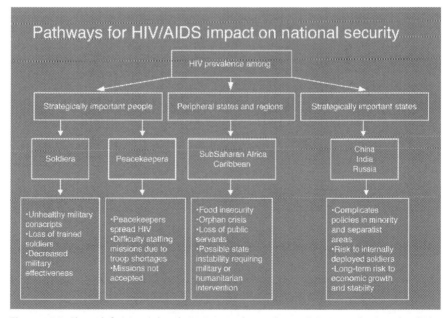

Figure 6.9 Shared foreign, development and security policies emphasize the links between development, stability, and epidemic control.

References

[1] Kevany S, Sahak O, Workneh N, Saeedzai A. Global health diplomacy investments in Afghanistan: adaptations and outcomes of Global Fund malaria programs. Med Confl Surv 2014;30(1):37–55.

[2] National Armies for Global health? Editorial, The Lancet, October 25, 2014.

[3] Kevany S. James Bond and Global Health Diplomacy. Int J Health Policy Manag 2015;4(x):1–4 Editorial commentary.

[4] Burkle FM. Throwing the baby out with ther bathwater: Can the military's role in global health crises be redeemed? Prehosp Disaster Med 2013;28(3):197–9.

[5] Novotny T, Adams V. Global health diplomacy—A call for a new field of teaching and research. San Francisco Med 2007;80(3):22–3.

[6] Center for Strategic and International Studies. Final report of the CSIS Commission on smart global health policy, 2010.

[7] Center for Strategic and International Studies. CSIS commission on smart power, 2007.

[8] Nye J. The decline of America's soft power. Foreign Affairs, 2004, May and June.

[9] Kevany S, Baker M. Applying smart power via Global Health Engagement. Joint Forces Quarterly 2016; 83, 4th Quarter.

[10] Feldbaum H, Michaud J. Health diplomacy and the enduring relevance of foreign policy interests. PLoS Med 2010;7(4):e1000226.

[11] Kevany S. International Access and Global health Diplomacy in Sudan. Lancet Glob Health 2014.

[12] Kevany S. Global health engagement in diplomacy, intelligence and counterterrorism: a system of standards. J Policing, Intell Count Terrorism 2016;11(1):84–92.

CHAPTER SEVEN

Trade-offs: ethics versus economics in epidemics

Abstract

The trade-offs between the ethical and the economic are a key feature of debates around resource allocation in infectious disease epidemic programs. In many cases, more cost-effective interventions are prioritized over those that are more expensive (such as HIV/AIDS treatment versus prevention). However, global health diplomacy arguments and evaluation techniques suggest the need to look beyond utilitarian considerations, and take more holistic views about the hidden benign and malign effects of health interventions at both health and nonhealth levels. Using examples from field experiences, the ways in which ethical and economic arguments can be reconciled through barefoot diplomacy are explored.

7.1 Numbers and realities

Global health is rooted in the real—and yet in infectious disease control, theoretical ethics and economics are essential (Fig. 7.1). The abstract arbiters of debate, discussion, and decision making; the catalysts for controversy and change. Ethics and economics thus guide us between the Scylla and Charybdis of cost effectiveness and utilitarianism through exposing their limitations: barefoot diplomacy instead reconciles and equilibrates, by creating the programs that satisfy both ethicists and economists.

In any infectious disease effort, ethics and economics determine distributions of effort across projects and programs, in various regions and populations. Life-altering decisions are informed by cost-effectiveness analyses in conjunction with humanity and instinct—by predicting how many lives will be saved or how many infections will be averted, measured per dollar spent. Decisions are also informed by resource allocation scenarios—through comparative portrayals of health results against money spent.

Economics therefore governs achievement of optimal results: but economics, in turn, is governed by ethics. But the ethics of humanitarianism is the apogee of the utilitarian; ethics and economics therefore coexist in a state of tension: a representative of trade-offs between, say, treatment and prevention. Policymakers and program managers have to make trade-offs

Barefoot Global Health Diplomacy.
DOI: https://doi.org/10.1016/B978-0-12-818681-7.00006-6

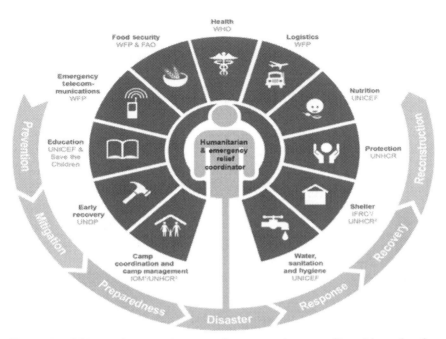

Figure 7.1 Ethics and economics may often come in to conflict with each other amidst the vast array of considerations required for global health efforts.

between spending on programs to save a single life in front of your eyes—or programs to prevent a thousand new cases of disease, far away.

Perhaps there is something inhumane about such trade-offs. These are impossible predicaments, that seem to demand another way. But alternatives demand a challenge to assumptions that our resources have to be limited. Can the assumption of too-scarce resources be trumped by diplomatic arguments? By the diplomatic results of infectious disease control, which can generate enough support to treat and prevent—so that it's not necessary to choose?

Barefoot diplomacy is a challenge to utilitarianism—it undermines logic, and challenges measurable programmatic virtue, through *realpolitik*. Utilitarianism is sometimes akin to a right angle in nature—the economics doesn't always jive with reality [1]. Utilitarianism in the form of spreadsheets and numbers is theoretical, numerical effectiveness; however, spreadsheets and numbers are disconnected, at times, from cultures, environments, settings, religions, and societies: from the local utilization and acceptance of services.

Disconnected also at times from the downstream: from inadvertent smart effects, that can only be seen out of the corner of one's eye. Functionalism can miss the combined effects of medicine and diplomacy—the blurred results between humanitarianism and peace, aid and trade. Conversely, such dual effects can generate new rationales for supporting infectious disease control efforts—which in turn, challenges perceived limitations on the budgets and resources that demand trade-offs between the ethical and the economic.

Cost-effectiveness is too often indifferent to diplomacy—and yet, cost-effectiveness has been in the ascendancy. Utilitarianism, one could argue, is a fashion, a system of belief: a by-product, perhaps, of the tech era. A consequence of calculators and computers, working out every cost associated with every program: through working out the costs and the outcomes, the machines eventually decide what programs are valuable, based on the numbers and the budgets. Yet there are constant difficulties in comparing different costs in different places, which include difficulties in measurement: include also issues of changing metrics and prices, inflation and results, and generating consumable, transient truths [2].

An ideal world is perhaps one with fewer disposable truths, and greater epidemic response budgets; a world that ensures that every local who needs treatment for an infectious disease gets it—without net loss to the donors (Fig. 7.2). There are millions of people around the world dying from HIV/AIDS,

Figure 7.2 In rural Zimbabwe, local conditions require that epidemic control efforts be both ethical and affordable, as most income is required for basic needs. *Picture courtesy Sebastian Kevany.*

malaria, or tuberculosis: in an ideal world, they would be treated and cured, regardless of the cost. Yet, so often, the cost is seen as too high in comparison to the returns: the short-term economic trumps the ethical in the collective, global economic logic and consciousness world. The results of investment, as we have seen seem inadequate when they are based only on narrow, altruistic returns—when results are measured only in terms of health outcomes.

With barefoot diplomacy approaches, the health outcomes are just the tip of the iceberg. Health is just one part of a range of global goods which are produced through international relations, prestige, and humanitarianism; smart programs can produce trade and security dividends through enlightened self-interest, cooperation, and conflict resolution—all, beyond the grasp of utilitarianism's narrow measures. But recognition of the potential spectrum of local and global goods produced by combining health and diplomacy programs is challenging the old beliefs and paradigms; challenging 20th century decision-making systems, and the necessity of trade-offs they are predicated on [3].

In HIV/AIDS, for example, prevention is cost-effective—and yet prevention is also an invisible effect. It overlooks the diplomatic cachet of treatment and rescue: the argument of tangible results, and compelling recoveries. The latter casts epidemic control—along with its supporters and advocates, acolytes and providers—as a compassionate and diplomatic movement. Yet program decisions are often made purely on the basis of cost-effectiveness—which requires, too often, elements of humanity and compassion to be sacrificed. At the extreme, cost-effectiveness requires different values to be placed on different lives to determine who should be saved: invidious different values on age and education, background and productivity. Patient prospects, wealth, and value to society can end up informing both choices and decisions: utilitarianism demands categorizations of the value of a life.

Through smarter structures, which generates more compelling cases for greater support, there is the potential to both achieve better decision-making systems and avoid such unpleasant analyses and decisions. Cases and arguments can be made that ensure there is enough health for everyone: to create a world in which more support and investment becomes available for health programmes, not least because of their downstream effects. As we have seen, if you spend on health in smart ways—on the right programs, in the right places, at the right times—other troubles will be resolved as well, dispensing the need for separate investments. Global health diplomacy is thus creating a case for more support, both ethically and economically; a case that everyone understands and connects with. A case coherent—economically and ethically—to voters and politicians, taxpayers and businessmen, the political left and right.

Barefoot diplomacy is also challenging health economists—echoing past rejections of cost-effectiveness decision-making in affluent lands [4]. Past experiments with utilitarian approaches are, perhaps, now being exposed as too robotic, mechanistic, and insensitive to the personal. Two examples from my experience spring to mind: first, the often-inappropriate cultural context of male circumcision programs for HIV/AIDS, which—though lauded as being cost-effective—had little or no awareness of the challenges of introducing routine surgery throughout sub-Saharan Africa. Second, the reproductive health interventions I came across in Kenya [5] that were, because of the depth and breadth of intervention choices made available to poorly educated locals, too often mistrusted by locals who saw them as efforts to eradicate reproductive capacity.

We are challenging judgements that have to be made, sometimes, on a human basis: by doctors, nurses, and health workers for the patients in front of them; by those trained in the Hippocratic tradition to offer the best possible treatment. These judgements aspire to the individual and ethical—as well as the social and utilitarian—optimum: [6] it is the distribution of resources best for the patient, as well as best for society, as Mcliza Flores describes in the Philippine context.

Health care in the Philippines

Health care inequality in the Philippines is complex. Primarily, who you know, what you have and where you live is the divide. If you live in the urban areas such as Makati, Quezon City, and Metro Manila you have access to the best hospitals in the country such as Medical City, St. Luke's and Makati Medical Center. These are a few of the privately owned hospitals with state-of-the-art equipment and facilities. Doctors at these hospitals are highly competent-mostly Filipinos who received their educations from Ivy League schools around the world, and bring excellent qualifications. However, the cost is quite absurd. Like any other businesses, making a fortune is the goal.

During one of my visits in the home country, my father needed medical attention. Private hospitals have ambulance services, but with the traffic congestion, it is faster to drive the patient to the hospital. On the way, a call was put to his primary doctor. By the time we arrived at the ER, the said doctor was there to personally attend to my father. The physicians and medical staff were very professional, thorough and attentive. The cost may be absurd—but it is worth knowing that you get high-quality care, service and attention when it is a life or death situation.

(Continued)

(Continued)

But what about the ones who cannot afford the cost? There are government funded hospitals: you will find trained medical professionals in these places but the lack of funds associated with corruption and politics, equipment and medicine makes providing quality health-care challenging.

The rural and remote provincial areas suffer the most in disease burden because of inequities in access to services. They have to travel longer distances to get to the smaller hospitals or clinics, with very little medical supplies, lack of equipment and no available qualified health providers because of the salary factor. I have a niece that attended medical school: during the course, a group of them volunteered for a medical mission in one of the remote areas. They were able to raise the funds, and gathered medical supplies with the donations and support from private sectors that believed in their cause. Humbled, she expressed to me that they could have done more to help the people but that such goals were unattainable due to limited supplies and funds.

Major food-borne, water borne and mosquito-borne infectious diseases pose considerable health threats in these areas, and the population have very little or no experience of getting treatments. The ones who are meek and needy—to see a doctor is close to a dream—even with the government's free health-care program, need the money that they do not have to travel the distance. They are thus often left with no other recourse but rely on the local medicine person that performs traditional and spiritual healing.

One thing, for sure, that is not difficult to understand and hasn't changed is that in all aspects of life, the poor suffers the most. If you are poor in the Philippines, you have no connection with known public figures and no economic strata; you are most likely to die. Finding the equilibrium to all these inequalities will require a miracle...

Meliza Flores

7.2 Beyond mere quantification

Cost-effectiveness systems, in epidemic contexts, are also occasional excuses for cutting costs—for streamlining, rationalizing. They are excuses for cuts to programs during eras of altruistic contraction. Yet, there is still a demand for the same amount of health, for less cost; for the best of both worlds. Rationalizing, streamlining, can therefore be good—until it goes too far; until society loses more than it expects. Losing programs that, though less cost-effective, perform on other levels—such as the diplomatic, the humanitarian. Losing programs that perform on levels of security: identifying the downstream effects that generate greater support; the forgotten trade-offs, the invisible variables.

Such trade-offs—as well as those between the social and the individual, the ethical and the economic—are represented by the Adam Smith Problem [7]. The father of modern economics described trade-offs between compassion and self-interest; decided that there is no room for altruism in economics. He described the latter as incompatible with utilitarianism: as an unnatural, counterintuitive element of human behavior—unless there is a *quid pro quo*.

Under Smith's invisible hand, all resource allocation decisions, in theory, are optimally-made based on cost-effectiveness. But in the real world—in the land of *realpolitik*—the Adam Smith Problem rarely occurs. The real world, for once, is maybe better than the theoretical: a world in which benign considerations beyond theoretical ideals sometimes prevail. Considerations of adaptability, diplomacy, humanitarianism: considerations that often occur not to distant international scientists or health economists, but to local doctors and nurses. Locals, with a different outlook—beyond budgets, post economic. The locals—maybe ironically—sometimes seeing the big picture that the internationals miss.

The locals see lives saved with less cost-effective drugs. They thus see some success—even though it's not always the streamlined, economist's dream. And they see, in front of them., ostensibly less efficient programs that build better relations with communities, or that advance conflict resolution. That work in ways that capture attention and imagination; in ways that attract political and public support. In ways that resonate with generals as much as with generalists who might, otherwise, proceed without an understanding or appreciation of epidemic control; and who might even consider it a waste of money.

Enlightened locals and internationals—by combining ethical, economic, diplomatic arguments to sell global health across political, social spectrums—can thus win support for programs that are cost-effective—as well as ones that, sometimes, are not. They can win over the skeptical, the indifferent, those who are critical of infections disease control or public health emergency responses. Barefoot diplomacy efforts can field skeptical dismissiveness, disdain, criticism: can respond to critics of investment in infectious disease programs in poor places in ways that leave doubters doubting themselves.

Support for infectious disease control across spectrums thus allows the achievement of both individual and social optimums; allows the reconciliation of economics, ethics, and the Rule of Rescue. The latter posits that effective medical resources should be used, as and when they are available, to help the sick; as opposed to the social ethical optimum, which maximizes the net benefits of health expenditure across all members of society—even if individual sacrifices are required [8].

Ethical and economic conflicts in epidemic contexts in the 21st century can be reconciled through diplomatic styles and outcomes - through new ideas and systems. New structures, that generate better allocations of scarce medical resources without abandoning the sick in the interests of efficiency. Structures that reject, both ethically and diplomatically, the off-limits; the analyses that recommend withholding life-saving treatment when it is available and affordable. Reconciliation of philosophical perspectives is therefore possible through added investment, which in turn depends on new arguments for infectious disease control investment and support.

These arguments make the world, us, want to invest in epidemic and infectious disease control; they illuminate returns, and challenge the inevitability of trade-offs; they build instead on human creativity and on alternative solutions (Fig. 7.3). These arguments deconstruct paradigms, question assumptions and constraints: constraints of limited budgets and resources for epidemic or pandemic responses (once, passively accepted) can now be challenged by arguments that remind decision makers that evaluation and support of health services is not only a technical matter, but a quintessentially ethical endeavor. In complex societies with diverse values, there are, in essence, a range of considerations that trump the utilitarianism and rationality of cost effectiveness analysis [9].

So, barefoot diplomacy looks beyond utilitarianism: it generates alternative rationales for how programs should be designed, delivered and funded. It considers not only narrow measures of success—the numbers of cholera infections prevented, malaria bed nets distributed, tuberculosis cases identified—but, also whether those programs have a culture of measurement and accountability. It considers whether services are being used, and whether aggregated results and achievements are being communicated by internationals to locals; considers whether the program is adaptable, or whether it responds to local as well as international pre-determined priorities. These are the new questions asked of epidemic response efforts, which go beyond cost-effectiveness: new philosophical perspectives that broaden the range of qualities required for the gold standard epidemic control effort.

Figure 7.3 The daily allowance in hyperinflation Zimbabwe: barely enough to buy basic medical needs. *Picture courtesy Sebastian Kevany.*

Such altruism is thus moving from the neo-utilitarian to the Rawlsian: [10] to approaches combining the ethical, diplomatic, and economic. It is casting the net of values ever wider: does the program advance equity, or social justice? Does it build dignity, through confidence and self-worth amongst communities and individuals, patients and locals? Does the program function in terms of sustainability, transferability, future commitments; in terms of visibility, on the local and international stage?

From Rawlsian values to Kantian; to theories of perpetual peace: [11] the search continues in this era, as in every previous era, to discover new solutions to world conflict. Solutions that can maybe be advanced through the innovative evaluation of medical interventions—by values assigned to program effects beyond health results: what is the value of a contribution of a tuberculosis program to nation building, peace keeping, regional stability, or conflict resolution; does an epidemic control program in Southern Sudan also improve coordination between international partners and countries, or abet succession efforts? As Stuart Garret illustrates in our next essay, not all significant medical efforts can be deemed cost-effective.

A journey, grasshoppers and a blessing: progress of the rehabilitation and inclusion agenda in rural Uganda

A common exercise in development training is to ask individuals about the Sustainable Development Goals (SDGs) they most identify with. For me these are SDGs 3 (good health and well-being), 4 (quality education) and 17 (partnerships for the goals). As a founder and leader of an Irish registered charity focused on "developing healthcare together" by sending professional health care volunteers overseas to LMICs, this alignment seems reasonable.

Having recently completed an MBA I now understand that defining and sustaining a consistent strategy can be challenging in the charity sector: a charity is established to deliver a social impact for those it is set up to support. Adding value for shareholders, in the context of a charity, is achieved by adding social value for stakeholders. Charities must utilize their resources in the most efficient and responsible way possible, in order to maximize the benefit for those they support. For charities, impact drives income and income drives greater capacity to support your social mission, creating a virtuous cycle. Charities are, therefore, competing for a limited amount of resources (income or funding) to add value and impact to their social mission. In the charity sector, I believe the aim is to build a sustainable impact. For me, developing a strategy for a sustainable impact that aligned to SDGs 3, 4, and 17 began with a journey on a motorbike in Uganda.

In 2011, I was working as a physiotherapist at a rural field hospital called Kisiizi Hospital in south-west Uganda. On this particular day, Ugandan Occupational Therapist Batringaya Alozious asked me to visit a patient with him for a joint physiotherapy and occupational therapy assessment in Kanungu—a village that was an unknown distance from the hospital. Alozious could not ride a *peche peche* (motorbike), and I had only passed my motorbike driving test 3 weeks prior. What Alozious did not know was that I had passed my motorbike test and secured my license in Dublin (where roads are flat, and traffic lights, road signs and markings ubiquitous). Despite being qualified to drive in Ireland, driving on the beautifully uneven rich red roads of rural Uganda was more like driving on another planet! Adding Alozious as a passenger made it even more challenging. Before setting off I asked Alozious just how far Kanugu was from the hospital; he vaguely replied "Not far—maybe like 1 hour."

We set off at midday on the small one litre Indian made Boxer Baja, laden with water and the equipment required to complete our assessments. Some two hours, later we were still driving. The rich red roads had become narrow mountain paths and we were climbing higher and higher into a progressively thickening mist. We then came to a path that descended downward to a four inch plank of wood crossing a ravine with no other way round it. I looked at

(Continued)

(Continued)

Alozious and he nodded towards the plank of wood and then proceeded to get off the motorbike and walk across first. Grateful for the confidence (or lack thereof) that Alozious had shown in me, I slowly let the bike roll down the remaining hill and kept the front wheel dead center, proceeding to cross the plank and to my surprise, safely reaching the other side.

We continued on, and eventually the tea plantation at Kanungu loomed out of the mist. We were now on the border with the Congo, a good 3 and half hours from the hospital; we were greeted with an exuberant welcome from a family who had not been expecting a *Mzungu* (white man) to attend. We assessed Sincere, a cheerful 6-year-old girl with Cerebral Palsy. We made some recommendations, taught Kashilling Alice (Sincere's grandmother and guardian) some stretches and exercises, and measured Sincere for some supportive equipment.

While this was going on, the clouds had become increasingly darker and now looked ominous. Conscious of the time, the darkening sky and my limited driving skills (yet to be tested at night) I turned to Alozious and said I think we should leave *Zooba Zooba* (quickly). However, Sincere's family had other plans: two plates of hot grasshoppers, a delicacy in Uganda, were presented to us. Alozious said we had to eat them as a sign of respect. I forced down the warm salty insects, and wondered if this day could get any more interesting!

The sky outside was becoming darker still, and I reminded Alozious once again that we should leave as the rain was about to come. As we stepped outside of the house, Alice and two other family members presented us with a large sack of Irish potatoes, a fully-grown white hen, and a small basket made from woven grass. Alice said they were gifts from Sincere, and they wished us a safe journey. We thanked Alice, and posed for a photograph. My colleague Alozious was quick to accept the gifts, and began strapping the large sack of potatoes to the passenger seat of the motorbike just as the rain began to fall heavily. Within minutes rivers of red muddy water began to flow where the road had been. I looked at Alozious and asked him what we would do now? He laughed, and said *"We ride!"*

I then looked hesitantly at the hen tucked under his arm. Alozious exclaimed: "No—we must take the chicken." I started the motorbike, and we jumped on. I struggled to even get the motorcycle up the hill from Sincere's house; visibility was limited, with the mist growing thicker and the rain heavier. The hen clucked loudly as we hit large wet mounds of red mud, and Alozious laughed nervously as I struggled to keep the bike upright. We then reached a section of road that was completely water-logged, while a yellow digger was attempting to clear it. I drove with my legs out like stabilizers to try and get us through. However, the back of the bike continually slipped out

(Continued)

(Continued)

from under us, and we repeatedly fell into the quagmire. I told Alozious (who was still holding his chicken) to walk through the 100 m section of muddy road while I walked, pushed, pulled and drove the motorbike and the sack of potatoes through. It was a sight to behold—and many people gathered to watch from the shelter of their corrugated iron covered porches. The white Irish man toiling through the mud with his sack of potatoes and motorbike must have made for an entertaining scene.

We eventually got down off the mountain, reached a flat piece of road, and stopped for a short call (toilet break) at a small village. After we had relieved ourselves, wet and now cold, we began our journey again. However, the motorbike would not start. More entertainment for the locals, who gathered to watch. No one was offering to help, as it was still lashing rain and the show was too good to interrupt!

Eventually, I got the bike started and we headed back to the hospital. We were almost home when we came to a large steep section of road. It had been raining so heavily that the water had created large potholes in the steep slope. I had to drive slowly around these potholes, much like a skier negotiating a steep slope—complete with moguls to add to the challenge! I negotiated the first three potholes, but the bike slid over the edge of the fourth and we went with it, falling off the bike and rolling down the hill. I slid to the bottom with the bike; Alozious managed to stop himself halfway, but his chicken was not so fortunate. As her legs had been tied together she proceeded to roll, flap and cluck loudly the whole way down the slope. We checked nothing was broken, made sure Alozious's chicken was still alive, and drove gingerly back to the hospital saturated and covered in mud. My pride of successfully securing my motorbike license some 3 weeks prior had now lost in the red wet mud of Uganda.

This day, though, was important. We had completed our assessment of Sincere, provided exercises and stretches to her grandmother, and given the measurements to the carpenter to make the supportive equipment. We had learned from her grandmother Alice that she believed in Sincere, and was doing her best to put her through school with her brothers and sisters. By providing us with the grass hoppers, potatoes, chicken and small basket Alice had cleverly (and by tradition in Uganda) locked us into a partnership—thus SDG 3, 4, and 17 for me were being contributed to, while for her, she instilled a vision in us. This vision is essential when we know *there is a substantial and ever-increasing unmet need for rehabilitation worldwide, which is particularly profound in LMICs*" (WHO, 2018) and *"the availability of accessible and affordable rehabilitation plays a fundamental role in achieving SDG 3"* (UN, 2015).

(Continued)

(Continued)

Figure 7.4 Stu and Alozious before their memorable return trip. *Courtesy of Stu Garrett.*

Roll forward some 6 years, and sincere is 12-years-old. She has attended three Cerebral Palsy camps, established and run by teams of Ugandan and Irish health-care professionals and volunteers. She has continuous donor support from Ireland, and has been assessed by a novel project at the Kisiizi Hospital primary school (which our charity established in partnership with the school management, the local rehabilitation department, an orphan's sponsorship program and a charity run by special needs teachers in the UK). Since the program's inception in 2016, it has provided mainstream primary school education and a rehabilitation service to 14 children with disabilities, with two children progressing on to second level education.

This is a significant achievement, given the poverty, local perceptions of disability, and the communities understanding of special needs. The small organization that I now lead supports with rehabilitation assessments, equipment and funding for teachers in special needs. We have also sourced an Irish secondary school to partner with the program, and they provide funding for the additional needs of the children that the other stakeholders in the

(Continued)

(Continued)

partnership do not contribute to. The local primary school and the hospital now also have a long-term vision of a special needs class room linked with the hospital's rehabilitation service.

To date, my focus has been to lead my organization on the journey of developing and supporting this novel and challenging partnership. There have been many bumps in the road, literally and figuratively, but every time we have fallen off I have ensured we have gotten back on and continued the journey, as I strongly believe we are contributing to SDGs 3, 4, and 17. Despite the challenges, I am committed to this program and to making a sustainable impact. For sure, it will be a harder journey than the wet, muddy, treacherous motorbike adventure when I first met Sincere—but in Uganda, when it rains, it is considered a blessing!

Stu Garrett

7.3 McNamara fallacies

The evolution of philosophical approaches to global health thus offers a broadening of what to consider, what to assess (Fig. 7.4). It is an evolution that bridges self-interest and altruism; that leverages enlightened self-interest as a unifying force. Above all, it is an evolution that widens the range of outcomes and considerations that are used to judge an infectious disease program's value; an evolution that recognizes the reality of both self-interested and enlightened incentives for support, involvement, and investment based on a deepening understanding of evaluation of the unobserved outcomes and effects.

Yet the dramatic outcomes of barefoot diplomacy beyond health are often never captured. Nonrecognition of these outcomes is resonant of the McNamara Fallacy, [12] of Vietnam-era statistics. This fallacy, named for Robert McNamara, the US secretary of defense from 1961 to 1968, is based on the temptation to view the world and to make decisions that are based solely on what can easily be measured. During the Vietnam War, McNamara's measurement were of villages conquered, prisoners captured, and areas secured.

But, as history reminds us, statistics only told half of the story—the numbers that disregarded all that couldn't be easily measured. Figures that overlooked the power of ideology, climate, or local knowledge: information that misled and governed flawed approaches, based on narrow evaluations.

Systems that were not dissimilar from the narrow, economic appraisal of today's epidemic response programs; that marginalized the immeasurable. Numbers in war and in global health can therefore taunt us—even if you can't measure the effect, it doesn't mean that it doesn't exist.

McNamara Fallacies, in war as much as in health, are the epitome of the self-deception of misinformation. They represent a belief that you get not reality but what you measure. This is wrongheaded in the health context as much as in fields of conflict because the invisible effects—the diplomatic, or world-stability dividends—still exist, whether measured or not: some public health emergency response programs create better international relations, or make contributions to humanitarianism and human rights, security and diplomacy. Some programs are better at producing those effects than others; perhaps those are the ones to back; the projects that mobilize and support imagination, and that generate more resources for health, thus satisfying ethicists and economists, skeptics and believers.

An example: in South Africa, amongst both skeptics and believers, I worked in townships on HIV/AIDS programs that didn't always measure what was produced. At the time, public health responses were divisive, political issues; linked to advocacy, protest and social unrest. it all wove into debates on pharmaceutical profits, or generic drug production; lapped with ethical debates on access to life-saving treatment campaigns by catalyzing legal and cultural progressiveness. The situation was also intertwined with debates such as legalizing prostitution or gay marriage in order to destigmatize and save lives: debates which connected with the new South Africa paradigm, as Madiba called for more support for care of the poor. Together, these debates connected South Africa to the world: all were involved in that tragic, thrilling moment in time.

It was exciting to be there—to be involved, even on the edge. The edge of a change—of a groundswell against a utilitarianism that frowned on treatment; that challenged accepted standards and the *status quo*. A change also that generated an outbreak of interest and compassion in the lives of the damned: being there, I saw tensions over support for treatment programs resolved, magically, by awareness of the downstream. By statesmen and policymakers around the world tuning into the invisible benefits of treatment—outside logical cause-and-effect equations—that resulted in many lives saved.

Lives saved, in distant clinics and forgotten hospitals, because of the recognition of epidemic response effects that included altruism and humanity, security and prestige (Fig. 7.5). Lives were saved, equally,

Figure 7.5 In Southern Africa, though equitable and diplomatic, getting to remote rural sites, requiring treks cross-country, can rarely be described as cost-effective. *Picture courtesy Sebastian Kevany.*

because of political and diplomatic outcomes; because of politicians and diplomats catching a glimpse of enlightened self-interest. Seeing benefits to themselves and their countries, as much as to the sick and dying, by their glimpsing of economic, international relations, and regional stability benefits of health programs: a fleeting vision of altruism's true reward. And so politicians and diplomats did what the political, the diplomatic world does best: tap in to the zeitgeist, and leverage.

Thus diplomats and doctors, politicians and nurses, advanced universal antiretroviral treatment for HIV/AIDS as well its benign domino effects—but there were still problems, all kinds of problems. Problems of expanded treatment for some patients coming at the cost of other health-care services; of neglecting those conditions attracting less visibility, attention, or prestige [6]. There were problems, as well, in challenging established utilitarian ideology: problems of distant scientists saying—scientifically and economically, epidemiologically and technically—that money spent on life-saving drugs was a waste. These scientists, with perhaps narrow visions, were possibly missing the wood for the trees—missing, maybe, the cachet of drugs that saved lives. The *je ne said quoi* that was catalyzing support—financial and emotional—amongst locals and internationals; support that would otherwise not exist.

Health economists in ivory towers, with arguments as airtight as their offices: such economists were oblivious to bigger pictures; isolated from smart designs, and cool effects. Such economists were overlooking ideals even more transcendent than optimizing resources; ignoring dreams of more support for epidemic control shared by locals and internationals; of more infectious disease control for poor people, in poor places. They were perhaps overlooking, as well, support for arguments and polices, programs and plans, that achieved goals not by redirecting money and support from the North or the West—not by redistributing—but by augmenting. Missing arguments and rationales that tapped into resources beyond worldwide health expenditure—resources previously used elsewhere. Resources earmarked, maybe, for diplomacy or trade—because, even if they were imperfect, antiretroviral programs were producing those domino effects as well.

At the time, the downstream effects were only visible through instinct and peripheral vision: the effects were, as yet, uncharted. Only recognized by our deep, intuitive, human logi; the recognition of hidden effects was a first step, maybe, against the dominance of utilitarianism—of cost effectiveness, and robotic logic. Nothing more than a vague recognition of the invisibles, of the intangibles—of the downstream, the human. The hidden effects, that were consequent on the glamour and heroism of antiretroviral treatment; consequent on health and nonhealth effects, that captured the popular imagination. Elsewhere in South Africa, the effects of the epidemic were being felt by those such as our next essayist, Emily Avera.

Empathetic teaching and learning in South African blood services

"I want to just ask them why did they donate blood when they know that they have HIV?"—Jamie, research nurse, in a tone of frustration

"I felt like they were judging me."—Carlos, a regular blood and platelet donor, recounting an instance when he disclosed having a new sexual partner (considered a risk factor for HIV and other STI's) and was thus deferred from donation.

As these quotations from my fieldwork with the South African National Blood Services (SANBS) staff and donors reveal, the impulse to pass judgment is all too common. Jamie, after qualitative research methods training,

(Continued)

(Continued)

eventually learned what was valuable about withholding judgment in interviewing donors, but I also wonder if her earlier moral indignation which I quote above was a similar to the sentiments of the nurse who reviewed Carlos' health and lifestyle questionnaire at a blood drive. Juxtaposing their experiences compels me to think about what could be learned when we lean into empathy instead of judgment, and what is foreclosed in the impulse to judge.

While the empathetic consideration of many perspectives—rather than, say, an "objectively" wrought data point—in making sense of any given problem or question is integral in any qualitative and ethnographic research methodology, the opportunity to develop training and to teach methods has been a healthy reminder how much discipline and hard work it takes to hone a multi-perspectival sensibility. In my experience working with blood services' personnel and people in the health sector more broadly in South Africa, adopting a multi-perspectival or more open-ended approach to research is often a huge step outside their comfort zone—one that they regularly treat with skepticism. However, if people are willing to make an effort to embrace these approaches and take their validity seriously, there are both a variety of rewarding outcomes and challenges in undertaking such an endeavor.

Along with Tara, another anthropologist based at a university in California, I was tasked with training SANBS research nurses in ethnographic semi-structured qualitative interviewing. The goal was to equip these nurses with the ability to both recruit and conduct interviews with people who had attempted to donate blood but tested positive for HIV and were already on antiretroviral (ARV) therapy. The question at hand was similar to Jamie's, though not inflected with the same judgmental tone: why did these people attempt to donate blood when they had presumably been diagnosed and receiving treatment for HIV?

It was unlikely such a question would be answered through quantitative methods that the research team was already familiar with. The supervising doctor, Anneke, asserted that they needed qualitative research to better understand this conundrum: accepting presumptive answers to the research question and simply testing for these would shut down a whole host of potential answers and reasons.

The sensitive and complicated circumstances surrounding people's situations, the potential contradictions of their rationales and their emotions, could not be captured as effectively otherwise. Nor would any policies or practices derived from the research be as efficacious. While Anneke was not an expert in the methodologies Tara and I were teaching, and she was well-versed in quantitative methods, she was a ready advocate for the value of qualitative

(Continued)

(Continued)

approaches in the blood services among her colleagues. Without her cross-disciplinary diplomacy, the qualitative methods would have been more difficult to teach, and the discomfort and judgmental attitudes about qualitative research prevalent in the biomedical realm would have been much more difficult to overcome.

In the midst of all of these various aspects of training, however, there were two lessons in the interactional politics of interviewing that we found very difficult to teach: (1) refrain from passing judgment on the participant, and (2) ask open-ended questions that let the participant show you what is significant for them. Why were these things so difficult for the research team to learn?

Part of it was an initial suspicion that they already had with respect to the potential participants in the study. From their perspective, early on, it was inconceivable and morally reprehensible for these participants to think of trying to donate blood if they were knowingly HIV positive. Did they not think of the risk that this posed for the patient receiving that unit of blood? Or had they not considered the other smaller, and yet still significant risk that a wayward venepuncture poses a risk of infection to the phlebotomist or other people at a blood drive? Or did they think they were cured because they were on ARVs, or did they not think they were positive anymore now that they were being treated? These kinds of suspicions, and their desire to satisfy their own assumptions about the scope of reasons for the participants' actions, were very difficult to set aside or surmount.

As a co-instructors, Tara and I needed to maintain an empathetic stance in addressing these difficult lessons. The nurses' initial judgmental attitude is a product of the nature of their work in SANBS, whose operations are centered on the prevention and mitigation of the risk posed by a donor transmitting an infection to a patient. For better or worse, risk management and safety are paramount in blood service work, not just between nurses and donors but in practically every department of the blood services. The vigilance around risk as the foremost concern in their work is completely understandable to me, but the primacy of this orientation would seriously constrain the empathy and openness required for the purposes of the research at hand.

Since their training, the research team has continued to grow in the capacity to be open, set aside judgment, and balance the demands of semi-structured interviewing. I am excited to see how their research, now in progress, turns out. In short, not only does the general work of research in the health sector—or any sector, really—require the *skills* of empathic engagement, but effective teaching and learning requires a similar engagement too.

Emily Avera

7.4 Reconciling ethics and economics in epidemics

Barefoot diplomacy can thus sidestep trade-offs between tricky decisions. It sidesteps, as we have seem through new perspectives; through nullifying decisions between treatment and prevention. It sidesteps also through aspiring towards ideal worlds in which constraints on support, money, and budgets—and the trade-offs they demand—are challenged. Such constraints are overcome by joint ethical and economic, altruistic and self-interested response programs that allow for both optimal treatment and prevention. Old constraints and choices are thus made irrelevant by another way; a third way, that recognizes the security, cooperation, and humanitarian consequences of greater public health support.

A third way, also, that recognizes infectious disease control improving the lives of the haves as well as the have-nots: that recognizes that the worth of epidemic response health investments is demonstrated only partially by altruistic effects through better health for the poor. The worth of a malaria program, when locals and internationals are smart about it, is demonstrated alongside better international relations: it enhances its worth through equity and conflict resolution; peace and trade, productivity and growth. Enhanced human security, improved through the synergistic effects of both health and diplomacy. But the dual worth that satisfies both economists and ethicists, and that generates wider returns on investment, only comes from the right program, in the right place, at the right time.

Such dual worth is also indicative of the limits of science and technology: it recognizes their critical role, their brilliance, but also their limitations. Wider perspectives change old systems and standards of program design, by recognizing that ephemeral effects aren't always scientific; cannot always be quantified. Instead, optimizing health outcomes in both ethical and economical, human and computerized ways, pays off in terms of narrow outcomes but also via ripple and domino effects. Through inclusion of such principles in program design and delivery, assessment and evaluation, there are diplomatic, security, trade and environmental repercussions optimizing them, and thereby broadening global health's appeal and support will broaden investment, in turn savings more lives.

Yet optimizing altruism's downstream effects requires an inherent awareness of bigger pictures: without their consideration, the opportunities to achieve multiple goals including the health and the social, the political and the developmental, may be lost [13]. We thus require recognition of a world which is becoming more and more intertwined, and less distinct: require acceptance of a world in which every line is blurred; where old divisions between disciplines and responsibilities, endeavors and styles, are being broken down. A recognition of a world in which efforts to build barriers and to isolate disciplines is increasingly exposed as futile; as attempts to defy the inevitable.

Lines between self-interest and altruism, between the theories of Adam Smith and the realities of human behavior, also need to be blurred (Fig. 7.6). As do lines between local and international concerns; between what is best for the local, and best for the global. Barefoot diplomacy, in that context, could be imagined as a Venn diagram, in which realms of local priorities and international prerogatives coincide in which the circles overlap, in a zone of enlightened self-interest [14].

A zone, further, that makes the case for global health investment: that argues for it ethically and economically, diplomatically and compassionately. A zone, also, that engages governments and politicians—the left and the right—in newly compelling ways: ways that demonstrate that smart, cool programs can also benefit the national interest; the environmental and political, security and economic.

Diplomatically designed global health, epidemic response, and public health emergency programs and their effects, ultimately will enhance their demand and supply, locally and internationally, and amongst both the haves and have-nots. In economics jargon, it will shift the demand curve for response programs through recognition of their broader benefits—in keeping with economic theory—via changes in expectations of what they produce [15]. Smarter efforts will thereby enhance program functionality and productivity—enhancing value in the same way that smart phones can be more valuable than land lines. In the same way that the world, wisely and intuitively, invests more in tools and systems that can do more—faster, better, smarter. There is more value, as well, in efforts that heals the sick but also resolve conflict, saves the environment, creates work, or improves security. Generating something for everyone is also making the case for altruism based on diplomatic, ethical, economic, as well as health impacts.

Perhaps we can, then, ultimately reconcile the ethical and the economic, allowing optimization of both via greater support and more resources. As a global community, we can do this by meeting the demands of activists and policy makers—of ministries of finance and

Figure 7.6 In rural Tanzania, it is only through extensive local consultation that ethical and economic considerations can be reconciled. *Picture courtesy Sebastian Kevany.*

health, populists and progressives, right and left. Reconcile also by meeting the demands of nurses in distant jungle clinics, and health economists in modern office blocks: by harmonizing the demands of locals and internationals—each, often, with seemingly irreconcilable perspectives; each with different priorities and world views, in which altruism may or may not play a part. Those initiatives that save lives as well as improve security, cooperation, diplomacy, the environment, and the economy are hard to reject.

And the augmented support generated by such programs is already saving lives. Smart instincts have already led to support for epidemic response treatment programs that are improving the health of millions. I've seen them save lives—in spite of utilitarian arguments against them— around the world, because of the invisible, enlightened ripples that such care creates. Seen programs that met the needs of diplomats and statesmen, hawks and doves; that met the expectations and demands of epidemiologists and doctors, skeptics and the general population—that demonstrate that there is something in it for them, as well.

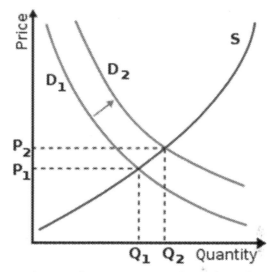

Figure 7.7 The complexities of economics mean that shifts in demand curves based on ethical considerations can drive demand for more expensive, and les cost-effective, services.

7.5 Key messages

- The ethics of epidemic control can often come in to conflict with economic and budgetary considerations on aid and assistance.
- However, attempting to illustrate diplomatic and security dividends from infectious disease control can help to address this dilemma.
- With demonstration of the dual effects of smart programs, the case for greater health budgets worldwide is greatly strengthened.
- In turn, this results in the possibility of providing health care or epidemic control interventions that are both economical and ethical (Fig. 7.7).

References

[1] Kevany S, Matthews M. Diplomacy and health? The end of the utilitarian era. Int J Health Policy Manag 2017.
[2] Kevany S, Benatar S, Fleischer T. Improving resource allocations for health and HIV in South Africa: bioethical, cost-effectiveness and health diplomacy considerations. Glob Public Health 2013;8:570—87.
[3] Ibid.
[4] Royal Institute for International Affairs. Global health diplomacy: a way forward in international affairs. Chatham House: meeting summary document; 2010.

[5] Shade Starley B, Kevany Sebastian, Onono Maricianah, Ochieng George, Steinfeld Rachel L, Grossman Daniel, et al. Cost, cost-efficiency and cost-effectiveness of integrated family planning and HIV services. Aids 2013;27:S87—92.

[6] Fleischer T, Kevany S, Benatar S. Will escalating spending on HIV treatment displace funding for other diseases? South Afr Med J 2010;100(1):32—4.

[7] Nieli R. Spheres of intimacy and the Adam Smith problem. J Hist Ideas 1986.

[8] Cookson R, McCabe A, Tsuchiya A. Public healthcare resource allocation and the rule of rescue. J Med Ethics 2007;34:540—4.

[9] Benatar SR, Gill S, Bakker I. Global health and the global economic crisis. Am J Public Health 2011;101:646—53.

[10] Rawls J. A theory of justice, (1971). Cambridge, MA: Belknap Press of Harvard University Press; 1999.

[11] Kant E. Theory of Perpetual Peace 1888.

[12] Basler M. Utility of the McNamara Fallacy. Br Med J 2009;339 b3141.23.

[13] Katz R, Kornblet S, Arnold G, Lief E, Fischer J. Defining health diplomacy: changing demands in the era of globalization. Milbank Q 2011;89:503—23.

[14] de Tocqueville A. Democracy in America 1865. In: Kassalow J, editor. Why health is important to U.S. foreign policy. New York: Council of Foreign Relations; 2001.

[15] Grossman M. On the concept of health capital and the demand for health. J Political Econ 1972.

CHAPTER EIGHT

Blurring the line: a review of barefoot global health diplomacy

Abstract

Global health is increasingly concerned with nonmedical issues. These include, but are not limited to, conflict resolution, counter terrorism, environment, other international development objectives, diplomacy, international relations, the foreign policy of donors, and the holistic well-being of the planet. In this context, we review concepts related to barefoot diplomacy, which are focused on the interaction or interdigitation between infectious diseases, primary health care, maternal and reproductive health, malaria, other epidemic control, and health system strengthening and these concerns. For many trained in clinical sciences or public health these are subject areas which may be unfamiliar; efforts are made to review these concepts in user-friendly ways and demonstrate how global health can impact nonhealth outcomes in both malign and benign ways.

8.1 Linking global health with diplomacy redux

We end, as we began, with Kennedy's question: do concerns of war and peace; of security and justice; of international relations and diplomacy, really permeate every walk of life? Should they? If so, how can each of us—in any fractional way—do something, do anything, to advance these concerns? If everything is linked in one way or another—every job, life, action, and decision makes a contribution to the global quotient of peace, stability and cooperation, then how can we tune in to that?

If so, what unexpected magic can we create as a result? Can the right program, or epidemic response, or public health emergency protocol—one which is usually perhaps concerned with caring for lower-income populations—also help the environment? Can the right response program, delivered in the right place at the right time, also moderate extremism, alleviate poverty and inequality, resolve territorial disputes? Can it even end wars, and advance world peace?

Smart, cool, magic, renaissance—such terms are incongruous with infectious disease. Smart—like a smartphone, something that does more than you expect, that operates on more than one level at once without threatening the functionality of its *raison d'être*. Cool—as in fashionable, hip, attractive, stylish,

Barefoot Global Health Diplomacy.
DOI: https://doi.org/10.1016/B978-0-12-818681-7.00004-2
175

appropriate, moderating. Magic—as in the effects of the unexpected and often immeasurable; renaissance, and in the rebirth of a paradigm in new and creative ways. Such are the innovations that our world now calls for, both as a way of spurring investment in the world's health and also ensuring that related responses advance both health and nonhealth agendas (Fig. 8.1).

Global health could be considered as a catchy term in today's world most would consider it to be a positive idea. Important to the world: to rich and poor; donors and recipients; locals and internationals. Global health (which was in many ways formerly and alternatively known as public health, international health, community health, or world health) is usually associated as dealing with the infectious diseases, epidemics, related living conditions, and health services of the poorest people in the poorest regions of the world. This concept expands down dirt roads to smoky villages; through jungles; along the ever-expanding deserts; from the international sphere down to the local, individual level. Yet, as we have found, it also exists on everyone's doorstep.

So, looking back over this book—what is infectious disease control all about? It is generally perceived to be about treating and preventing, tracking

Figure 8.1 A news article in a South Sudanese newspaper. *Picture courtesy Sebastian Kevany.*

and testing, for epidemics. It is about building up health systems—including building actual structures, and providing equipment, referrals, education, and access. It is about everything required to provide health care to those who need it most; it is about helping the forgotten. It is about inequality and poverty, environment and climate change, global warming versus trade and conflict; about communications, travel, defense, and security.

It is about arguments regarding globalization and migration. It is about the political left and right; about economics, politics, diplomacy, terrorism, cooperation, money, and the distribution of resources; about materialism, consumerism, capitalism, communism. Medical interventions are also about progress, war, peace, conflict, spirituality, politics, and international relations: It is about hearts and minds, alliances and allegiances, forging ententes and détentes; about the future of the world [1].

Global health thus touches everyone in a manner similar to war and peace: just like Kennedy said. However, not everyone may agree with every idea. These are divided times: consensus is hard to come by. Many people believe—and some passionately believe—that diplomacy, epidemics, world peace, and the environment should all be addressed separately. Many believe that there is no place in altruism for anything but the functional; the pure, the focused. Or, at the opposite extreme, many believe that altruism is a luxury, a waste, a folly: "Kennedy's dream"; a hopeless effort to bridge irreconcilable gaps, to unify polarized perspectives. The impossible dream: a distant vision of renaissance, consensus, enlightenment.

An impossible dream for so long; but, all these years later, one that perhaps may now be possible. Today, infectious disease control is a critical part of the universal machine of humanity; part of the international apparatus, the new world order, the 21st century life. It is represented on the world stage by organizations and individuals with power, reach, influence—within and beyond health,—medical versions of FIFA[1] in football. And so epidemic and pandemic responses become politicized, popularized, diplomatized. It is an "Allegory of the Cave"[2]; whereby the ostensible

---

[1] *Fédération Internationale de Football Association*: Football's, or soccer's, world governing body.

[2] Plato's exploration of the deceptiveness of face value. According to Wikipedia, *"Plato has Socrates describe a gathering of people who have lived chained to the wall of a cave all of their lives, facing a blank wall. The people watch shadows projected on the wall from things passing in front of a fire behind them, and they begin to give names to these shadows. The shadows are as close as the prisoners get to viewing reality. He then explains how the philosopher is like a prisoner who is freed from the cave and comes to understand that the shadows on the wall do not make up reality at all, for he can perceive the true form of reality rather than the mere shadows seen by the prisoners".*

effects of responses to epidemics or PHEs only just scratch the surface of the true impact and reach of such altruism.

Global health was once a subset of foreign policies and affairs—thought of as esoteric, obscure—not much of a scene. However, not any more—with the rise of epidemic risk awareness, as well as global responses, concerns, exposures, and communications. Nor can it be thought of as separate entity with the rise of supra-national health organizations (with or without national flags), with mandates to augment awareness of global goods (Fig. 8.2).

Nor can infectious disease control be separated from globalism and globalization, or with our new societal aspirations and dreams and values: the concerns and priorities of the 21st century, of the digital age. We now have our new, globalized, transcendent ideals: Likewise, global health is now a bigger concept—because it is compelling, urgent, tangible. It is big because it also has to do with drama, tragedy, success; with heroism, rescue, and humanity. And, subtly, big because it has to do with wider effects; deeper consequences of conflict and international relations, on the world stage.

Figure 8.2 Sometimes, epidemic control missions can feel like trips to other planets, requiring all one's nous and *savoir-faire*. Remote South Sudan. *Picture courtesy Sebastian Kevany.*

Still, not everyone is keen. Infectious disease control has been known to carry the label of a "misplaced priority," a waste of money. "What is the return on our investment?" is a common question in this sphere. Why are there free international handouts, when there are sick people at home? What are we getting back for investing in the health of poor people far away—when we have problems, and issues of our own?

Sometimes, the skeptics and doubters have a point. There is a lot of dodgy idealism within the global health field, along with a lot of waste and expense on the money merry-go-round. Many would gladly see it all dramatically and radically cut back; allow it to get dropped from the agenda. The counterpoise argues for the support—to defend such investments, while also trying to optimize the direct and indirect effects. Defending, by demonstrating that there are magical effects of infectious disease control programs which go far beyond health—if you can somehow capture them, or work out which programs deliver them.

But how—by having to optimize, not only with numbers of people who are given life-saving drugs in distant villages, but also the downstream, ephemeral, nebulous elements: the knock-on effects, the dominos, and ripple results? The conflict resolution, equality, growth dividends; the increased cooperation, better communications, security, environmentalism. The nation-building and peacekeeping, justice—and social justice—and, of course, the good international relations vibes that go hand in hand with epidemic responses as well.

Global health, like anything, has to sing for its supper—from refuting doubters, to convincing skeptics, to finessing isolationists. It needs to demonstrate, for example, that through HIV/AIDS or other epidemic prevention and control programmes a lot of death, misery, conflict and sickness has been avoided—both in donor and recipient countries. That, as a result, for a lot of very, very poor people there is access to a lot better health. That, even if more investment is needed—more programs and drugs and support—what is already in place also has to be defended and justified. And the one overarching rationale—the one key motivator, one central catalyst—for doing this is to try to articulate the subtle, broader effects epidemic control or public health emergency efforts can have beyond health, on the economy, on the stability of the world. To manage and to optimize and to credit those effects; thereby to have infectious disease control understood as doing more—much more—than meets the eye.

Epidemic response efforts thus have to change as the world around us changes. Every day, there is an increasing overlap between domains and purviews: between security and stability, economics and health, leisure and work, defense and cooperation. Between trade, military; business, diplomacy. Boundaries are getting blurred, while functions are becoming multifarious. Every part of society, in these "smart" times we live in, is being asked—or rather demanded—to synergize, to streamline. This is perhaps a reflection of the world we live in—a world where excessively narrow focuses, functionalism and utilitarianism can rob us of the beauty of that which is indirectly—and sometimes accidentally—produced [2].

Thus the smart world we live in now, in which it has become increasingly hard, almost impossible, to silo or separate endeavors. It is now hard to isolate epidemic responses from other concerns; harder and harder to fence off disciplines and occupations, jobs or ideologies, from the ripples they create. No effort on the world stage, these days, is insulated from the effects it causes. The domino effects: capturing, troubleshooting, and ensuring these downstream effects requires new, renaissance, holistic perspectives; requires new styles, and a new consciousness. Without this, as Dr. Travis Bias describes, the end no longer justifies the means.

The "Global Health" Martyr

The year 2000 brought a wave of health programs answering the call of the Millennium Development Goals. The early campaigns took aim at a single disease, such as HIV/AIDS. The international goals for addressing all factors that affect epidemics, including the social determinants, have since matured to prompt efforts to strengthen entire health systems, widening their scope to corral badly needed resources to also combat noncommunicable diseases that cause the majority of deaths worldwide.

The programs to address these Sustainable Development Goals have exploded, created and supported by nongovernmental organizations (NGOs), governments, and universities. The workforce deployed, often by high-income countries to low-income settings, can involve researchers, medical students, nurses, or physicians—to name just a few. Yet no matter the years of experience, even the best trained health worker sometimes does not fully appreciate the health system, societal, or political context in which they are about to work. This can have a detrimental effect on the worker, on the host system or university, and on wider diplomatic relationships between programs—and even countries.

I have accumulated experience working and teaching in East Africa, in both rural and urban settings, in NGO and government-funded initiatives, alongside academics, physicians, medical students, and medical residents. These posts have introduced me to, and made apparent the need to confront, one of the most disturbing personas I have come across: the global health martyr.

(Continued)

(Continued)
The global health Martyr

I arrived on the packed open-air medical ward to meet my new colleague. He was exasperated. There was a medical ward full of 72 patients to be seen, and the medical officer interns were nowhere to be found. There had been three deaths overnight, and he had not received a single phone call. The nurses had his local mobile number—why hadn't they called?! He had been there in the Kenyan country for a year. His well-rounded training as a physician in the US had prepared him well to take care of patients in any setting, but his own physical signs of burnout were apparent. Seemingly on the verge of tears, he was defeated—not just by the daily preventable loss of life, but by the exhaustion of staying on call every single night since he had arrived last summer. If he wasn't going to show up to help these patients, who would?

In a neighboring country, a new wave of health volunteers arrived from the States. A group of red-eyed nurses, physicians, and midwives grabbed their luggage, and jumped into vans waiting to take them to the hotel. They were welcomed by one of the nurse volunteers from the same program who had already been in the country for two years. She would serve as their main host, and the one tasked with orienting them to their new culture for the next year. Over instant coffee the next morning, she told them about the work she had undertaken during her time of service.

The volunteers were also welcomed by various Ministry leaders and oriented by their program's staff for two weeks, before separating into the different towns where they had been assigned. But the nurse volunteer gradually appeared reluctant to collaborate with newly arrived colleagues at her institution: she seemed hesitant to share materials from previous projects, or even PowerPoint slides she had used for lectures. It became apparent she had not built positive relationships with other local NGO workers, or even ward nurses. In fact, it sounded like she was working on projects strikingly similar to those from other organizations in town, but she did not seem to be aware of their work.

The small pet projects of hers looked great to the American organization that sent her, but no one else knew much of the details of her successes or failures. What *had* she accomplished over the past two years? Only she knew, and in fact, that seemed to be her goal: if she was indispensable, she could secure funding to keep her in the country another year. She was not eager to return home to work in the US. In each of these locations, a desire for service, glamorized by *The Last King of Scotland* scenarios—or worse, an older, more paternalistic colonialist model—had transformed, perhaps unconsciously, into a self-serving and destructive mission [3].

(*Continued*)

(Continued)
The Martyr, defined

Historically, a martyr is one who gives up their life for a cause, often with religious underpinnings [4]. Their belief in a dogma bigger than themselves drives them to suffer and to make the ultimate sacrifice; sacrificing their life to further the quest for many others. In health care, the metaphorical martyr has been glorified. Displaying signs of martyrdom—the cranky surgeon barking orders in the operating room; the fatigued nurse working 4 night shifts in a row; the drained medical resident who has been on call every three nights for a month—has traditionally been revered by superiors, commanding a badge of courage.

But society is shifting, and this behavior is increasingly becoming unacceptable. Many have come to openly recognize that it is unsafe to be exhausted while taking care of patients: medical resident work hours now have restrictions. And it is not okay to shout at teammates in the hospital; verbally abusive colleagues are being reported and held to account.

If the martyr is not welcome at home, why then should expatriate health workers be allowed to export this model abroad? That mantra ought to be at the forefront of our minds whenever working outside of our home culture: if it is not okay at home, it should not be okay here.

The Martyr's impact

The first physician I met could feel himself breaking down. But he was unaware of the ripple effect emanating from his intense desire to help: he was raising his voice to nurses and even to his boss who was, as he saw it, unwilling to hold the interns accountable for showing up to work. During lectures and bedside teaching, his tone of voice discouraged soft-spoken (common per local custom) students from speaking up or asking questions; he was simultaneously losing credibility within the department and isolating himself.

The results were damaging to him personally, to the health system, and to diplomatic goals of his program—and even his country. Personally, he was suffering from exhaustion and demoralization, which were compounded by witnessing avertible death on a regular basis. Posttraumatic stress is often written about in military circles, but he was setting himself up for a challenging return home.

On the system level, his actions were harming the education of those around him as he abruptly corrected or dismissed students—and worse, his condition was preventing him from making sound patient care decisions; this was apparent to colleagues. His presence at rounds even seemed to allow the local attending physician to justify regular absences, potentially obstructing medical students from locally relevant teaching, and patients from locally appropriate care. Finally, at a higher level, he inevitably stood out as an expatriate: on top of obvious racial differences, stylistic differences and cultural insensitivity caused uncomfortable discussions about him

(Continued)

(Continued)

amongst local staff. Everyone at the hospital knew which program had sent him, and he was thus unavoidably a representative of his home country.

Perhaps this was just an unfortunate case of a servant ground down by limitations in their work. Poor accountability for local health workers to show up for work, local hospital corruption with regard to a new wing being built, and leadership failures within the Ministry of Health were largely foreign to him. This judgement, though, is driven by source country expectations, built up over lengthy medical or research training at home. The failure to understand local context and adapt is therefore a wider problem, faced potentially by anyone funded by efforts to reach international health goals with expatriate workers.

If not the Martyr, then who?

The global health martyr can also take the form of the self-important NGO representative who has carved out a little section of work that only that volunteer understands; that seemingly only that worker can carry out. This is inherently antithetical to commonly stated NGO goals of coordination amongst other parties carrying out similar work, thus at times creating confusion with local health staff. There is plenty of work to be done in these settings, and a true leader would be working to build on previous successes or knowledge in the area—to pass on tasks to local stakeholders, rather than the next generation of foreign servants. At the very least, sharing with potential local collaborators could prevent duplication of efforts, and further progress towards achieving stated health goals. If in a neo-colonialist fashion, cooperation failures serve as a barrier to what ought to be the ultimate global health goal: no longer relying on international workers. Unfortunately, this martyr can effectively erase any such potential collateral gains of a well-intentioned program.

Preventing the Martyr

Achieving diplomatic health goals worldwide will require a hard look inward at universities, in NGOs, and within governmental programs; at all three levels of the aforementioned implications. Within each volunteer, there must be an intrinsic motivation, but this must be harnessed for good and not allowed to overshadow the needs of the intended recipient of program benefits or donor resources. Health workers must be prepared for the context in which they will serve; be supported throughout their service with close feedback and access to psychological or other required services; and return with a repatriation plan. In fact, shorter-term volunteers may require even more cultural and contextual preparation to contribute safely, and with respect, in a brand new societal environment: the sending organization or funding entity must equip itself to monitor the performance of their volunteer, and ensure their resources are being used effectively. This might include frequent iterative evaluation

(Continued)

(Continued)

regarding the organization's fit, program appropriateness, and success in meeting the true stated needs of the population they wish to serve.

Finally, the diplomatic implications, for our country, our programs, and our institutions, must perhaps be considered more intentionally. Cuba has sent health workers abroad for several decades, with the intent of improving their stature in the eyes of the rest of the world [5]. The thought process is that if one has been taken care of by a Cuban physician or nurse, they will likely think favorably of all Cubans. This sort of approach can be to our benefit, or it can seriously backfire. If American health workers are seen as part of creating a hostile work environment, adding stress on top of that which already exists due to limited resources, how will local health workers view all future American (or other expatriate) health workers who come to work alongside them? If NGO workers are not fostering a collaborative, interdisciplinary approach employing productive communication tools, then how will health delivery programs in low-resource areas be encouraged to maximize efficiencies to eventually no longer require external assistance?

Unfortunately, toxic or undiplomatic personas can prevent our ability to partner effectively with key stakeholders and thwart a system's ability to grow its capacity to serve its population. When does the global health martyr serve as a barrier to progress rather than a catalyst for health?

Global health interventions are actually not unlike other forms of international involvement. Several decades ago, we armed Pakistani military partners and did not bolster that with a plan for the resulting regional security fallout. In 2003, we invaded Iraq first, then belatedly prepared for reconstruction and a multilateral approach to stability. In health diplomacy, we are deploying and repatriating workers built for service, but we are again lacking foresight—missing opportunities to truly partner with local colleagues and respond to their expressed needs as part of a larger strategy to sustainably improve their systems to a stage at which they longer need the expatriate health worker, as well as improve international relations.

The unwitting transformation of an idealistic, altruistic health worker into a martyr can be prevented. Every time the martyr emerges, however, we run the risk of torpedoing our progress towards reaching the goal of universal health care—and worse, perpetuating actions that are antagonistic to both external donors' and local governments' stated and broader goals. The goals themselves are laudable, but a toxic approach or representative can render a hospital, a clinic, or even a university, incapable of working towards the end of improved health for all [6].

Travis Bias

8.2 Beyond job descriptions

Anyone who has done it—who has been there, been involved—will know that, in the field, they are not only working on public health emergency responses, or epidemic control, or "global health." While out on the field, other things happen—and there are many different roles to play. More often than not, there are opportunities and demands which have nothing to do with the original reason why you thought you were there; nine times out of ten, there are opportunities to do more—infinitely more—than what is captured by numbers, metrics, measures.

There are opportunities to become holistic, connected and to be mindful of the full spectrum of the effects of our actions; chasing and achieving infectious disease control goals is enough, oftentimes more than enough, in itself—it is a beautiful and noble thing. But could there potentially be more than those narrow goals? With the evolution of the field, there is a possibility of nurturing a broader awareness of the broader world—the social, cultural, environmental, diplomatic—in everything we do.

Awareness, sensitivities, fields of vision—capacities for which those from more formal, professional backgrounds are not always renowned. The global health world is dominated by doctors, professors, doctorates: by medical people, viewing the world and their role in it, in that narrow framework and context. Global health is rightly governed by the mentality of health outcomes—right, as far as it goes. But there is a catch: whether as a doctor, a nurse, an international worker or as a volunteer adventurer overseas—when working on a program in a distant place, you are inevitably operating under broader, invisible themes. Inexorably, through these themes there are diplomatic-style responsibilities and consequences both affecting and being affected by their actions—whether they like it or not [7].

Being apolitical, which for the medical community is often of high importance—doesn't mean being undiplomatic: doesn't divest one of those responsibilities. Perhaps it may thus be better to be tuned in, to be turned on to the bigger picture: the ability to see what's really going on. There is thus the associated idea that saying one is "just" a narrow operator—just there to do your job, to follow orders—lies the classical, dark, divestment of responsibility. Isn't this a style that that begets an end-justifies-the-means mentality—a mentality that achieves narrow targets at any cost?

These days, having a broader awareness is becoming unavoidable—no matter how hard one tries. Inexorable, for both individuals and organizations. In Southern Sudan I recall a local in a far out place, saying to me—"There is no greater threat to world peace than bad epidemic control efforts." Because, according to him, of hard-nosed, functional utilitarianism; a lack of finesse, diplomacy, and broader vision. Because of bad attitudes arriving along with the men in suits; the hype and the hustle. "There is a need," she said, "For anyone—not just in global health, but in any role, from any background—to weigh up actions, decisions, and consequences from wider perspectives. Weigh up the way that each move we make—our values and consumption, materialism and location, attitudes and politics, economics and professions—contribute to, or harm, the big picture."

How can the world make the case for treatment for people dying from the effects of distant epidemics or public health emergencies? It is illogical, in one sense. For HIV/AIDS, antiretroviral treatment is one of the least cost-effective interventions—even with generic drugs, or with ongoing efforts to convince pharmaceutical companies to lower prices. Economically, the whole idea becomes questionable ... And yet, bizarrely, irrationally, it is an intervention that is pursued—and one to which international organizations have given money on an unprecedented scale.

But why—why is there the impulse to do the irrational? How to deconstruct it, and see the logic of it? What are the other benefits the world should, somehow, be getting from paying for drugs for HIV/AIDS patients in far-out places? How can we see the antiStalinism of it: that one life saved, is more than a statistic? How can we see the parts that don't show up as figures—the invisibles, that subconsciously drive us down that road? [8].

The other side of the argument—what if providing epidemic control drugs, or responding to distant public health emergencies, is advancing world cooperation and stability? What if it is bringing locals and internationals together, through a dramatic gesture of ostensibly expensive altruism—that might actually be worth every penny? What if the money perhaps wouldn't have been there otherwise—what if other, more cost-effective, efforts didn't capture the world's imagination in the same way? What if locals will never forget when they were helped by international aid, when they needed it most? All that is surely worth something: not just better health for those in need, but also the additional moonlight effects (Fig. 8.3).

Figure 8.3 Even barefoot diplomats have to put on ties, once in a while. The author, teaching global health diplomacy and security to developing country participants in Lagos, Nigeria. *Courtesy Shayanne Martin.*

I have spent time working on HIV/AIDS, tuberculosis malaria, and other epidemic treatment, prevention, and control programs—often the one being sent to places that others weren't up for. Each project was always in the eye of a public health or political or conflict storm—some of them still are. In Zimbabwe, amidst currency crises and oppressive regimes, economic meltdown and global adverserialism, I spent time working on voluntary counseling and testing for HIV/AIDS: in Southern Sudan, rolling out malaria and tuberculosis programs in no man's land. To Northern Bahr el Ghazaal, Western Equatoria, and beyond: to the edge of the earth. To places that were hard to get to, hard to supply, expensive to run. To locals and internationals who were paid extra to work in hardship postings: all irrational on paper. Inefficient—until the invisible effects of global health programs are seen to result in peacekeeping, nation-building.

My vudents, and medical residents. These posts have introduced me to, and made to revolutionary Egypt; to edgy Papua New Guinea. I had a foot in the door of far out places, wild events. And, eventually—after the experiences, the adventures, and misadventures—impressions of ideas and situations started to weave together. Not knowing what or why—but knowing, for sure, that the right programs which worked felt cool, felt smart. That they seemed to produce, either consciously or unconsciously domino effects; that bigger things were clicking [9].

Querencia, and other Landscapes

In this chaotic and conflicted affected world, there is much to be said for a place from which one's strength is drawn, or where one feels at home. A veritable "enchanted garden" where one's exquisite loneliness can be forgotten, and where the human spirit—resplendent in all its beauty and resilience—can truly shine.

This landscape (or landscapes, as the case may be) for me rest simply and yet ironically in worlds most have forgotten—Myanmar, Solomon Islands, Sierra Leone; the Democratic Republic of Congo, Palestine, Haiti, Jordan and Papua New Guinea. In these post conflict jungles and deserts, what has been most startling is the absolute strength and fragility to be found in those who keep on—keep on living, keep on smiling, and keep on loving. The *realpolitik*. The human condition in all its darkness and light is deeply humbling. It resonates with a deep passion for being; Kremmer beautifully captured it in his book *The Carpet Wars* when he said, "they are everywhere, these individuals of undaunted humankind, irrepressibly optimistic and proud."

Paradoxically, I'm often reminded that humanitarians are either mercenaries, missionaries or misfits. I have oft questioned which I am. In a world of "agile learning," I have decided that I am a mercenary. Mercenary about learning, that is. Thus, leaving for my first of countless (and still counting) humanitarian missions as a young nurse of 24—like the Tiffany's ad—I breathlessly awaited: a revealing discovery, a promise of adventure; a whispered romance, a question answered and a secret kept.

Yet with such a list of expectations, the greatest lesson I have learnt from these undaunted individuals? Love. Love has manifested as diplomacy, as forgiveness, as reconciliation, as acceptance, as a path forward. Love, in other landscapes: this is where I feel that I am home. If only I could use the same ardor in being happy as I discovered from many of my colleagues and our patients—and their oftentimes unintended yet erudite lessons on life, death and love.

So whether you are a Mick Jagger or a Madonna, a Ghandi or a Mother Teresa (I'm told in the humanitarian world you are more likely to encounter types akin to the former duo—again, not sure what that says about me!) I've learnt that the main distinction between us lies in our capacity not to be scared by the differences between us; in our capacities to acknowledge and embrace the silence between our fear and love, conflict and peace, destruction and resolution, darkness and light.

Nothing typifies this more than the poetry of a landscape (or a good hashtag). Many a fine poet has likened the love of country and place, of feeling home, to romantic love. One of my favorite Palestinian poets—Marmoud Darwish—sums it up perfectly in his collection, *The Butterfly's Burden*. My favorite lines:

(Continued)

(Continued)

*"We listen to what hidden longing for a mysterious street
is in us: I have my life over there
my life that caravans made and then went on their way,
and here I have my life as my bread's worth
and my questions about a destiny a passing present
tortures, and I have a beautiful chaotic tomorrow
...so let love be an unknown, and
The unknown a kind of love. How strange
To believe this and still love!"*

In yearning for these mysterious streets and the unknown to be a kind of love, I've learnt that life (like love) is full of imperfections—although it is how we perceive them that renders them imperfect or perfectly acceptable. What matters most is the grace with which we meet, head on, these challenges—and, yet, at the same time we are aware of what we are and are not willing to concede in life. I simply am. Full stop.

So, to find that other landscape, the one that brings longing and loving, and the melody at night, that is from where I will draw my strength and be home. In what other landscape will you, yourself, be at home?

Amy Gildea

8.3 Nontraditional skills

In some way, maybe those domino effects have been happening on and off in global health since day one. Perhaps they have already produced downstream triumphs and failures: times of hitting the bullseye, of being ahead of their time without realizing it. There have been health campaigns that have generated ceasefires in civil wars for polio vaccinations to be administered: there have been overlaps and spill-overs between infectious disease control and everything else, producing the best of both worlds.

But there have also been risky and unplanned consequences: conceptually, sailing close to the wind. One can't ignore the risks of failure: of losing out, getting it wrong, making other things worse, because the type of program chosen was either too smart—or perhaps not smart enough. Is it going to capture the imagination? Will it sidestep echoes of cultural supremacy, or the potential for confrontation? Will the program adapt to local needs—will it jive with both global concerns, and local priorities?

There are limits—limits of multifariousness, of dovetailing and weaving agendas. Many question intelligence agencies using epidemic response programs to track down terrorists, if that leads to the death of innocent, uninvolved community health nurses? But, what if the terrorists hadn't been mere violent extremists but instead, another Hitler—a real life Bond villain, bent on killing, eradicating, conquering? What then? Would the trade-off then be acceptable? Global health can thus be coopted in to different definitions of altruism—but where is the threshold, the acceptable risk?

The need to balance and to reconcile different values in global health—all that is part of the riddle, as well. Most things in life come down to values. What you value—if anything—determining how you view the world: what you seek, what you see. Some value music, others travel; some work, others play. In the 20th century, different values didn't overlap as much, but these days, more and more, we have merging perspectives. At one stage, a bus company might never have cared how much a certain model polluted the world, as long as the vehicle got through its route on time. Now increasingly, the same bus companies are concerned about impact on the environment as well. More information, awareness and universal values—all working off each other.

Thus there are transcendent values for the 21st century: such values of the current world have to be taken into account in judging success or value of epidemic responses. In judging the success of infectious disease control, or of anything, are our efforts and endeavors responsive? Do they help or hurt the values of equity, social justice, human rights, or the environment? Along with transcendent values, there are now broader, renaissance standards of success: standards that apply not just to the pursuit of narrow goals, but to bigger pictures (Fig. 8.4).

But how? How do we manage to inculcate universal considerations? How do we decide if one epidemic control or public health emergency program is advancing human rights, or world peace, or the environment, more than another—and by how much? But to do that is to compellingly justify support for one program over another: to demonstrate that success in engaging national governments and the citizens of donor countries is the key; is where infectious disease control can be shown to benefit both the national and international interest [10]. That increased visibility will lead to greater funding [11]; that more evidence makes a more powerful case for the support of, say, public health emergency response investments [12].

Figure 8.4 It can be difficult to step in to diplomat mode when visiting infectious disease centers in remote areas of the Solomon Islands—particularly after long boat journeys. But the weather is some compensation. *Picture courtesy Sebastian Kevany.*

Such dreams and visions need new approaches, new ideas, to demonstrate and actualize the broader effectiveness of global health. An approach that generates holistic evidence of capacity across a world of interests, a universe of values. An idea for performance measures and program designs that win over hearts and minds of both locals and internationals; not just of the Global South, but equally of the Global North. An idea that wins over the "have-nots" and the "haves"—an idea that resonates across political, economic and social spectrums. From politicians and diplomats, statesmen and ambassadors, to the individual man or woman in the street.

Ideas that allow any background or profession to examine and optimize their own contributions to the *milieu* of epidemic control, conflict resolution, security, stability. Ideas and styles that ensure the best of both worlds—ways to maximize narrow results, while optimizing bigger pictures. Ways to spin the efforts and opportunities of infectious disease control into something much bigger. So, perhaps this is the story of how that can happen, in planned or unplanned ways—the story of the absurdities and excesses, antics and adventures, tragedies, intrigues, and *faux pas* those working in global health, around the world. Welcome to our stories of the tricks of the trade, and the ripples they create.

8.4 Intangible effects

In economics, changing expectations, roles and responsibilities are characterized by a shift in demand. Is this shift true for human evolution too? It is certainly true in the case of the demand for global health. These shifts of supply and demand represent responses to changes in taste: epidemic responses have been experiencing these shifts, with more demand from it on every level in recent years. So—while more is being supplied, more is paid for, more is invested; the equilibrium returns.

But what caused the shift—and how can it be maintained? Was it triggered by concerns with health security, by stopping epidemics such as Ebola from reaching home shores? Or was it catalyzed by exposure to the lives of the poor and the ultra-poor through our new world interconnectedness? Or is it just a result of changing human values, of advancing human sophistication, mobility, and evolution?

Either way, global health is now expected to deliver across a broader range of expectations and demands, on a greater number of levels, than ever before. Levels that include diplomacy; levels which can connect with political realms and deal with partnerships and ententes, resources and trade.

Diplomacy can connect with economics, with health—and with persuasion. But above all, diplomacy can be philosophically transcendent, and reflective of a kind of enlightened self-interest in which everyone gains. A brand of diplomacy that watches your back, so to speak—but does so in combination with the advancement and the protection of ideals. The transcendent ideals of human rights and democracy, governance and prevention of violent conflict, world security and stability.

Diplomacy, as we have seen, also plays in to hard power: the business of war and conflict. Hard power at its most extreme represents World War One, World War Two; but also, Iraq, Ukraine, Vietnam, Sarajevo, and more recently Syria. Hard power "is attained through military force, to influence the behavior and interests of other political bodies" [10]. Yet hard power and its effects are also forgettable, in the public mind: most people, in the developed world, never go to war zones. Most forget—or

Figure 8.5 In Sierra Leone, infectious disease efforts were combined with local surfing clubs via UNICEF support. *Picture courtesy Sebastian Kevany.*

cannot connect with—the challenges of them, any more than they can envisage the horrors of the distant past (Fig. 8.5).

Diplomacy—or smart power, in the military context—works by avoiding hard power where possible; by not using force to persuade. Instead, it seeks to use international relations, cultural exchange, development—and of course global health—as an adjunct. This kind of barefoot diplomacy is about building bridges: about solving local and global problems in a way that ideally avoids the use of ballistic power, yet that also draws on military resources and expertise.

To achieve these goals, defines and health organizations have to work together; both require diplomats. Diplomats are thus for preventing trouble; masters of the art of letting others have your own way.[3] To many, diplomats are a stereotype of suits, canapés, cocktails, and posh accents—helping relationships to work better between countries and individuals, locals and

---

[3] According to Wikipedia, attributable to Daniele Varè (1880–1956): *"an Italian expatriate diplomat and author, most famous for the China-set novel The Maker of Heavenly Trousers (1935) which was republished in 2012 by Penguin Modern Classics. He is also famous for the Laughing Diplomat (1938), his autobiography as Italian diplomat".*

internationals. Yet, good or bad, diplomacy is to most an exclusive art; a cloistered science. Today, the more traditional systems of diplomacy—the old ways, the ways of Talleyrands and Kissingers—seem cut off; rarefied.

Yet perhaps diplomats seem cut off because their styles are dated. Old-school diplomacy lacks the common touch: the informality, the mass appeal and input and style of the modern era. Put simply, the systems of diplomacy lack the cool of the millennials—of the day to day. Old-school diplomacy can be perceived as too shirty, passé, elitist: out of touch with the now: with democracy, transparency, accessibility, involvement. Diplomacy today needs a counter-culture and an offbeat style; it requires what I term "barefoot diplomats."

Fevers in the Jungle

I was, inappropriately, thrilled. In a way, I realized as I stared at the hotel room ceiling, I had been seeking these times—the intense tests of hardship in far out and hot and dusty places. The trials that would build a touch of strength, experience, karma: looking for the adventure, without doubt, but also the adversity.

Working on HIV/AIDS, tuberculosis, and malaria programs was rough and ready, I reflected as I stared at the mosquitoes on my wall—yet even allowing for the expected challenges (as the African love to call them), things had not worked out quite the way I expected. The levels of discomfort—just like the wild times and transcendent joys—were of a range and a variety that increasingly defied description.

The intensely unpalatable parts included all the basics and all the standard ones. Physical tiredness, jet lag, exhaustion; heat, cold, sensory overload, and the bewilderment that comes with new places. Included them all—but sometimes went beyond into another realm of disorientation: to waking up in nameless, placeless hotel rooms at night and not being able to figure out where I was, or even what country I was in. To having no way of figuring it out or even of recalling where the light switch or the window was: no choice but to lie there and wonder and wait to unthread the dream from the real; to decode the shadows on the wall.

I was in a state of global country-hopping disorientation in what had begun as a quixotic quest and that could only end with an element of exhaustion, I self-diagnosed as I stared at the ceiling. Borne of strange diets, atmospheres, climates; of getting on planes in summer and getting off in winter. Combine that with constant exposure to sickness and disease and epidemics and poverty—and then realize that is still just the tip of the iceberg. That there are the new faces, names, cultures, colors, and styles on the first day of every

(Continued)

(Continued)

mission—everyone else so keen and fresh, at the exact moments when you could o't keep jet-lagged eyes open. One was forced to be on the ball at moments when demands were highest but interest were lowest: there was, I decided as I got out of bed to see where I was, an element of disconnectedness that inevitably resulted.

Boundaries of time spent in such transient yet outré states of mind were, I reflected as I got up and felt around for the bathroom light switch, hard to define; hard to see where they started and ended. On some missions, the discomfort zones began instantly and lasted for days. Days when you get back to your hotel room in the late afternoon and lie on the bed with sunshine and noise, life and traffic outside and fall instantly asleep, fully clothed.

But by now, I knew was in too deep: a web of life themes now so entangled with the work, the existence: the pursuit of legacy and a professional evolution from cynicism to idealism. So many strange experiences, accumulating—and little choice but to enjoy it. Little choice, either, but to go down to the restaurant to get the last order in just before the kitchen closed at 10: finessing a meal, trying not to be the subject of the cooks' chagrin. No choice but to drink a glass of wine and go back to bed feeling even more discombobulated—no choice but to eventually, at 3 in the morning, give in to the jet lag that is yet again waiting in ambush.

Then, unexpectedly blindsiding you, the upswing. Through the grisly times and the discomfort zones like a plane through clouds; having lowered expectations, overnight the mission becomes so rewarding you don't want to leave: an organic rebalancing to having hit rock bottom on arrival, a counterpoise to the hotel-room lows.

The surprising releases and moments of relaxation, so easily forgotten: moments precipitated by talking football with the concierge: by a pink view, an orange sunset, a breeze from a green sea on a hotel balcony. By an anecdote or a new understanding or a new Swahili phrase—*hakuna matata*.

But first, the discomfort. Returning to bed from the bathroom, in a time zone 10 or 20 hours out, on the eve of a 2-week jaunt in to the jungle, I rise again at four in the morning to combat mosquitoes with my pillow. Battling them like a wartime general: by getting the easy ones (or were they the wise ones?) out by turning off all the lights and opening the door to the corridor so they flew towards the brighter glow outside. Getting them out, for sure, but having a fight to the death with the wily ones. I wondered how much my body could take: how long one could go without rest, as the ceiling fan lazily turned and I banished the last mosquito of the night.

But there still remained, in the silence after the buzzing had stopped, the echo of doubts—one's own, those of others, those of society. Each, layering

(Continued)

(Continued)

on top of the other like a symphony: that infectious disease control efforts are a waste of money; are ineffective. That they are bleeding heart escapades and do more harm than good; that they are inorganic and irrelevant and of marginal interest to the mainstream—that the Darwinist world didn't work like that, and has little time for exaggerated or self-inflicted tales of distress from far away places. Ultimately, idealism and belief could bring rest.

Despite all that, I would feel good—sublime, even—the next morning. The adrenaline of the new, kicking in via the simple pleasure of being given a strange African fruit for breakfast (or, even better, a shot of some bizarre local concoction).

Yet almost instantly the lurking discomfort was waiting to hit again: in a postprandial haze, I listened to myself agreeing passively to whatever timeframes and plans the local Ministry of Health or United Nations staffer was presenting to me in a hot, airless meeting room—they sensing my tiredness, no doubt, so presenting plans to me as a *fait accompli*.

And so, each time, I would agree to long and deliriously dehydrated drives on far-out trips to rural health clinics. To no rest, early starts, and seeing the effects of infectious disease in fly ridden tents in the middle of the jungle or township or desert. To heat and dust and jolts, and more and more insects: agreed each time readily, without energy for the necessities of bluff or negotiation.

Over time, as I learned the tricks of the trade or place, the challenge of combining global health and a sense of diplomacy would often seem to be getting easier. Yet it never really gets easier at all, because there is always a new place. Each time having to dial in to a new situation, environment, scene; vehicle, hotel bed. New laws, customs, culture; dangers, poisons, weather; noise, vibes, conflicts. Everywhere, also, the human and environmental questions that demand to know what on earth you were doing there and why you had come—but everywhere, each time and at some inevitable point, a level of total fulfilment, despite the echoing doubts, from one of the hardest, greatest jobs in the world.

Sebastian Kevany

8.5 Global stability and security

Of course, the world can't fix everything with diplomacy—any more than it can with infectious disease control. No hope of that, without interplay and combination—without barefoot diplomacy. Smart approaches—which improves the capacity of global health to attract and maintain funding, presents military-led alternatives to hard power interventions, and increases the prestige and importance of all stakeholders. Barefoot diplomacy thus appeals to these cosmopolitan values and inoculates, not just against infectious diseases, but also against hate, fear, conflict, extremism (REF). As renaissance paradigms, *ad hoc* diplomacy takes on soft power and shakes, stirs, and synergizes the field. Creates combinations that give it an edge—avant-garde partnerships which allow infectious disease control, or other manifestations of altruism and enlightenment, to do more—much more—than meets the eye. And doing more than meets the eye is the very epitome of barefoot diplomacy (Fig. 8.6).

So, in our response to epidemics and public health emergencies, why not break down barriers and blur lines? Who would not agree that, if renaissance-level design effects can be achieved, and if a spin-off program results in conflict resolution, or a better environment, or more cooperation, or less extremism, then there is more value and power in global health than might otherwise have been considered?

Who would not agree? Not only the soft power advocates, or the dreamers and the altruists. The cohort of neo-altruists that this idea is

Figure 8.6 Out in the field, the rough-night-s-before are always worth it. The author and friends, Solomon Islands.

designed to inspire might be handing out health, but they are not only giving hand-outs anymore: they are doing something greater. They are part of a catalytic combination—a progressive cocktail of backgrounds, interests, agendas, expertise and effects. Diplomacy and health and defense are beginning to work off each other and are producing, through new and interweaving roles, synergies with results that are more than the sum of their parts. Creating a new nexus: a new way of managing not only health, but also of managing the world, and helping to solve its problems.

But around the world, I have seen resistance to the nexus. Many put the blinkers on which are hiding them from bigger problems, bigger pictures—but I also was lucky enough to have met the dreamers, those who saw the bigger picture; the picture so big it transcended even health. The bigger picture of conflict resolution, of world stability and security: of directly or indirectly also pursuing and achieving those kind of results, along with epidemic responses. The dreamers reckoned that if an HIV/AIDS, malaria, tuberculosis, or other epidemic response program can do that, the work might tune into also doing something beneficial for a bigger purpose. Those who believe if programs can be designed so as to guarantee those downstream effects—every time, with no loss to health—then the world is on to an even better thing. Why? Because such perspectives in turn result in more money, more support, for the health of the world's poorest people, with everyone benefitting; the internationals as much as the locals, the haves as much as they have nots.

With most things in life, there is going to be an agenda. There is rarely pure altruism in the world—no matter how you slice it. But, perhaps there is an enlightened form of self-interest? There is, for example, global health efforts which is also enhancing the prestige of the donors, moderating the angst of former colonies, and improving world trade and security. Would the world still fund and support such efforts, without broader, nonhealth returns? Maybe—maybe not. As at the end of the day, in epidemic responses as in anything, there has to be a catalyst—an enlightened self-interest. So, what is that catalyst in today's terms? Is it our affluent and powerful countries recognizing the dangers of the radical inequality in the world? Seeing, as if for the first time, the lack of services and health care resulting in terrorists, fomenting revolution, and generating contagious diseases that will, ultimately, affect everyone equally?

Epidemic responses thus require not only treating the symptoms of world problems, but also the causes. Global health, in the 21st century, requires addressing other problems—without losing itself. It may result in health efforts affecting bigger pictures; not only acting as a palliative approaches. But there's also a need to measure and record, troubleshoot and optimize, these effects: to set up the systems that guarantee them. Systems that turn the occasional, the accidental and the serendipitous—the occasional improvement in national security through a public health emergency response—into the real, the consistent, the reliable (Fig. 8.7).

Perhaps there is a need for a system, a theory: how can altruistic efforts contribute beyond health? What if it can be proved to the world—to the dreamers and the skeptics, to the right and the left—that a clinic in Southern Sudan, if it is designed right, can achieve nation-building, security or peacekeeping effects? Would that make the case, simultaneously, for greater support of infectious disease control efforts? Would it make the case for militaries reinventing and repurposing to take on epidemic response or public health emergency roles—just as global health blurs its own lines, without sacrificing its *beau ideal*?

To try and argue that case requires tuning in to the real meaning behind the figures; the statistics that show the cost of one jet fighter can produce forty thousand village health centers. Some of that is sophistry, of course. The world has to have jet fighters, and perhaps

Figure 8.7 Combined nation-building and public health efforts in South Sudan.

better that some nations have more than others. You can't fight a justifiable war or defend territory with medicine or with diplomacy: the need will always exist for warriors. However, maybe today there is a more balanced set of needs and roles? Perhaps there is a threshold, an equilibrium—one that possibly was off-kilter in the 20th century, an era of overinvestment in ballistics, and always culminating in self-fulfilling prophecies.

Arguably, there are occasions when you can only achieve security by hard power. But there are also many times when such strategies have been exposed as ineffectual: there are times for iron fists, and times for velvet gloves. Times when the velvet glove works better—more efficiently, more effectively. Thus we return to the potential of global health's improved sensitivity to bigger issues, to diplomacy; of the field's nascent capacity to be able to see its potential to generate wider effects—beyond functionalism and utilitarianism, or other narrow outcomes. The smartest programs work by improving health service uptake through adaptability to local needs and cultures, tastes and styles (REF), which also win hearts and minds.

That is the up-side argument to making infectious disease control smarter. Yet this will require styles that are not everyone's cup of tea—there are still downsides, flipsides, pitfalls. Procedures, bureaucracy, the need to vet epidemic responses from broader strategic perspectives—these will all be spanners in the works. The world already has enough—in fact, more than enough—cross-checks and red tape. So—is it worth it—to make choices on programs beyond the cost-effective, the functional, the robotic? There are always going to be trade-offs: there are always going to be opportunity costs, lesser evils and greater goods. These are only acceptable if bigger pictures are advanced, as well.

The inexorable, uncomfortable bigger pictures: is a tuberculosis program aiding extremism, jihadism? Is it giving health to terrorists? Or is it the absence of an intervention—the lack of the right program, in the right place at the right time, and carried out in the right way—the root cause of related resentment and extremism? Is the real problem too little health—too much poverty? Is the problem the lingering economic and political values and beliefs of the 20th century, which empower angst and fear?

If so, is the solution to empower epidemic control efforts further through their people, their programs? To convince the general public to support and recognize the value of infectious disease control? To advocate

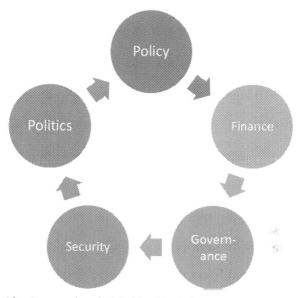

Figure 8.8 Epidemic control and global health diplomacy now involves a wide range of governance and societal functions, and relies on them all.

that hidden value to them; to involve them in those efforts? Let's be honest: global health is not an over-subscribed profession. It's not always glamorous, and most don't want the lifestyle, the thankless tasks even if they sometimes come with adventure, romance, and rare experiences. Most don't see that—until they step back a bit, and see that no effort is an island, and that epidemic control and public health emergency responses embrace and imply more than anyone could possibly realize (Fig. 8.8).

8.6 Key messages

- Barefoot global health diplomacy is ultimately about producing as many invisible positive downstream benign effects as possible.
- Without cultural considerations, as well as political and social, global health efforts may quickly lose support and momentum.

- Greater engagement with epidemic and infectious disease control will help to engage the global population on health levels more effectively.
- Other benefits of barefoot global health diplomacy approaches include effects on climate change, environment, security, stability, and human rights.
- There is a need for improved recognition of global health and epidemic control efforts to national and international security and stability.

References

[1] Sebastian K. Global health diplomacy, 'Smart Power', and the new world order. Glob Public Health: An Int J Res Policy Pract 2014. Available from: https://doi.org/10.1080/17441692.2014.921219.
[2] Center for Strategic and International Studies. (2010). Final report of the CSIS Commission on smart global health policy. Retrieved from: http://csis.org/event/rollout-final-report-csis-commission-smart-global-health-policy.
[3] IMDb. The Last King of Scotland; 2006. https://www.imdb.com/title/tt0455590/.
[4] Merriam-Webster. Martyr. https://www.merriam-webster.com/dictionary/martyr.
[5] Groll E. Cuba's greatest export? Medical diplomacy. Foreign Policy 2013; May 7.
[6] United Nations. About the sustainable development goals. https://www.un.org/sustainabledevelopment/sustainable-development-goals/. Accessed October 25, 2018.
[7] Farmer P, Kleinman A, Kim J, Basilico M. Reimagining global health: an introduction. Oakland, CA: University of California Press; 2013.
[8] Fleischer T, Kevany S, Benatar S. Will escalating spending on HIV treatment displace funding for other diseases? South Afr Med J 2010;100(1):32—4. Retrieved from: http://www.samj.org.za/index.php/samj/article/viewFile/3588/2650.
[9] Kevany S, Hatfield A, Beleke Y, Aurang Zeb B, White K, Workneh N. Diplomatic and operational adaptations to global health programs in post-conflict settings: Contributions of monitoring and evaluation systems to 'nation-building' in South Sudan. Med Confl Survival 2012;28:247—62. Retrieved from: http://www.ncbi.nlm.nih.gov/pubmed/23189590.
[10] Nye J. The decline of America's soft power. Foreign Aff 2004;.
[11] Katz R, Kornblet S, Arnold G, Lief E, Fischer J. Defining health diplomacy: changing demands in the era of globalization. Milbank Q 2011;89:503—23.
[12] Feldbaum H, Michaud J. Health diplomacy and the enduring relevance of foreign policy interests. PLoS Med 2010;7(4).

Health security considerations to improve the efficiency and cost-effectiveness of Ireland's future infectious disease and epidemic control efforts

Retrospect is easy, and Ireland's successes in epidemic control should not be understated: the shape of the country's mortality curve without interventions compared to what has been achieved would make for interesting viewing, were it possible to hypothesize such a counterfactual. But there nonetheless perhaps has to be a series of lessons learned from Ireland's mistakes—if they could be called that—in response to the epidemic threat. Like almost every other initially affected country (and far less than some), Ireland was unprepared in critical ways for onset.

Yet in future, such situations will inevitably occur under different circumstances; the world has changed dramatically since February of this year, in an unprecedentedly short space of time. Thus, what Ireland "should" have done then is no longer relevant—rather, Ireland's future health security may be determined by the application and implementation of a separate set of significantly more streamlined, controlled, and focused interventions that will ensure all the hard work (on both national and individual levels) so far is not undone.

Unfortunately, or fortunately, it is not just about stockpiling: while Ireland, as with much of the global north, was caught out in terms of medical supply preparedness, there is no way that this particular manifestation of an epidemic could have been realistically predicted in the public health procurement context. Future epidemics may also not necessarily be pulmonary or respiratory; while stockpiling is important—face mask and ICU preparedness, as well as the rapid conversion of other facilities to medical uses, should all be encouraged—such strategies are not enough in themselves, nor are they necessarily the most cost-effective options.

Borders and travel

Closing Air and Sea Ports: In terms of fast and effective—as well as cost-effective—responses, the immediate closure, or at least limitation on use, of air and sea ports, particularly from affected areas, might be a priority (within EU constructs). In retrospect this step, combined with enforced and monitored quarantine for repatriated citizens, would unquestionably have caused a significant reduction in infection rates. Of note, such efforts might also be combined with specific travel policies related to sporting events—as the disastrous consequences of the Cheltenham festival on Irish public health have illustrated.

Immediate Border, Air, and Sea Port Testing: For those ingress and egress pathways that remain open during lockdown periods, and preceding them, immediate port-of-entry testing should be prioritized. This might also be a consideration for the prioritization of roll-out of limited testing resources, with such locations taking precedence.

Priority Countries: A further international ingress and egress consideration might be the identification of priority countries based on the comprehensiveness of their response—though this may be complicated by stopover issues. Countries such as Belarus, for example, might be placed on highly limited ingress and egress policies, whereas those journeying from countries such as New Zealand or other nonaffected countries might enjoy greater travel flexibility.

Closing Borders: Determining an effective way to manage the Northern Ireland border in the health security context is clearly a priority. It is unlikely that Irish legislative influence will extend to Northern Ireland in the immediate future, or even the medium term; yet a cross-border working group on epidemic preparedness and policy cohesion may go some way towards addressing this.

Pass System Consideration: While it was vital for health services that public transport continue to function, enabling health-care workers to get to work, in many cases this leniency was abused by those making nonessential journeys. A pass system for essential staff to use on buses and trains might help to address this issue.

Adding Health Attaches to Embassy Staff: Many other countries, such as the United States and Norway, have made health attachés an integral part of international embassy staff. Their work—to transmit information on epidemic risks and assist with repatriation of epidemic-affected citizens, amongst many other day-to-day responsibilities including those related to aid programmes and HIV/AIDS efforts—helps the diplomatic corps to meaningfully contribute to public health efforts; Ireland may wish to consider initiating such a strategy forthwith.

Macro-level public health security policies

Short-Term, Intensive Responses: Too few restrictions scare, leaving society with a sense of fatalism and lack of control; Swedish approaches in this regard have been questioned in just the same way as American and British policies. Conversely, too many regulations, particularly over long time periods, are unsustainable. The solution might be to balance sustainability and reassurance: the imposition of more Draconian, immediate and wide-ranging restrictions—but for proportionately shorter time periods. This might include giving the Irish Defense Forces a significant initial role, rather than their recent provision of late-in-the-day support to overstretched law enforcement resources.

Immediate Geographical Distance Restrictions—Despite their Shortcomings: Mobility restrictions formed a key and highly effective element of the national response; for these to work comprehensively, however, they also need to be accompanied by hard borders beyond which no movement, except for urgent or emergency purposes, can take place. Otherwise, even two-kilometer limits can, in theory, form a spatial chain linking the entire country.

A Focus on Health-Care Worker and Vulnerable Population Safety: Future epidemics—should they resemble this one—will have the advantage of responses informed by previously identified risk factors such as age, comorbidity, and other considerations such as smoking or respiratory health issues. In all of those populations most affected in 2020, significant added protective measures should be allowed for—whether these are specifically for health-care workers, or tailored to the case of old-age retirement homes.

Contact Tracing: Contact tracing strategies have been described as "chasing shadows", and are indeed highly unlikely, in the age of media saturation, to have any real effect during the peak of a widespread and out-of-control epidemic. However, the use of tech to guide contact tracing strategies, even with all of the manifold problems that they face (and the frequent failure of such strategies in recent months) will undoubtedly help to form a further basis for early-stage epidemic containment in future public health efforts.

Law Enforcement Authority: The uncertainly and vagueness with which law enforcement was equipped with public health and epidemic control policing powers has caused extensive confusion. Only with a specific legal mandate to prosecute behaviors that threaten public health can the Gardai and others be expected to perform their duties with efficiency and accountability, should such events repeat themselves at any point in the future.

Varying Coping Capacity of Different Areas and Groups: For many population groups or other social demographics, recent months have amounted to inconvenience, fear, and frustration—but little more. Rather than blanket policies, far greater attention and support, in future, should perhaps be focused on those living in apartments or areas of high-density housing; those, also, who have been unable to access green spaces or beaches.

Virtual Schooling and Working: The speed with which Ireland successfully transitioned to home schooling and working could scarcely be improved upon; nonetheless, there exists an opportunity for more formalized contingency plans in this regard, covering supervision, curricula, and formal advice for parents and employers. With luck, many elements of the new status quo in this regard will persist, marking a much-needed paradigm shift in commuting (and therefore pollution and congestion) patterns.

Not Looking to the UK: Ireland has been, for so long, used to looking to the United Kingdom for leadership in global matters. Perhaps the recent epidemic has, along with Brexit, brought down the curtain once and for all on this era. UK policies have proved to be highly fallible; the groupthink of the European union, while also often flawed, might prove to be a far more effective paradigm in future.

Distinguishing Decision-Making from Advisory Boards: While centralized and authoritative control of national responses is often unfortunately essential in epidemic situations, the distinction between advisory and decision-making bodies should be considered. In the panic of the early stages of epidemic onset, unelected bodies without a formal decision-making remit may inadvertently become highly influential; their future role should be analyzed, not least to ensure that they do not eclipse pre-existing national emergency response entities.

Media and public health information

Immediate Release of Geographical Locations of Infections: Like so many other countries Ireland has discovered that, in public health emergencies, it risks becoming a victim of own bureaucracy. This consideration, linked to the loosening of restrictions on protected health information (PHI) in emergency situations, is of critical importance. Releasing locational information down to high levels of resolution and granularity—street level, if possible, but failing that district—will help the national effort on a number of levels: enabling those within this areas to take added care; enabling those outside those areas to avoid them; and enabling the government to focus both resources and restrictions on specific areas. Stigma risks are, perhaps, more than compensated for by reductions in fear, suspicion, and "false stigma".

Releasing PHI to Academics: One of the most problematic elements of the Irish response has been the dearth of analytical information (as opposed to the seemingly endless stream of commentary, expert or otherwise). For many epidemiologists, life in recent months has been business as usual, when their efforts might far more rewardingly, productively and valuably been focused on immediate initiation of research into risk factors, demographic data, and spatial epidemiology—to name but a few avenues of exploration. In future, it is perhaps important that the government consider allowing much faster data mining by Irelands academic community, with associated assistance in cutting the red tape and bureaucracy involved.

Surveillance and Knowledge Transfer: In many Ebola-affected countries, surveillance and knowledge transfer from field to centralized areas has been of particular importance. Establishing a civilian as well as a medical surveillance system for epidemic outbreak reporting may further help to ensure that future Irish responses are conducted in a targeted and focused manner.

Preventing Panic Hoarding: As previously observed, the necessary suddenness of the government's initial lockdown announcement was accompanied by much unnecessary fear, not least the undignified and demeaning scenes of panic buying and conflict at supermarkets during March. Fortunately, Irish society soon came to its senses, and has avoided the allure of hoarding (and therefore shortages); nonetheless, government public health messaging should perhaps include a specific "non-hoarding" component as well.

Broadcasters, Journalists and Bloggers are not Epidemiologists: The question of how to control the plethora of false theories, fake news, fearmongering, and paranoia generated by both the mainstream media and the internet is one that it almost impossible to address. Better, perhaps to advise society to direct its attention to a limited number of trustworthy news sources; over recent months, uninformed theories and speculation have abounded, leading to much unnecessary distress and fear. Too often, also, epidemic reporting has been conflated with political agendas: while many public figures have a role to play in terms of awareness building and messaging regarding the importance of adhering to government guide-lines, they should also perhaps be asked to recognize that they are rarely qualified to make informed opinions on specialist medical subjects.

Avoiding Excessive Concern with Economic Consequences: In an epidemic situation, there appear to be two economic options, neither of which are attractive: allow businesses to continue operating, even though their trade will inevitably suffer, and face higher death and infection rates—or suspend businesses, invest more in public health, and face similar sets of economic consequences. The key point is that, in both scenarios, economic losses are inevitable and unavoidable: those demanding more lassiez-faire approaches, or documenting the economic opportunity costs of restrictions, would do well to temper their criticism of government actions in this context. Indeed, there is a case to be made for suspending all such distracting and distressing economic impact analyses at such times.

A Health Benefit Focus: The closure of pubs may have led to many health benefits; similarly, two of the main risk factors associated with the epidemic are cigarettes and alcohol: the former because of behavior change and loosening inhibitions, as well as close confinements in small spaces; the latter because of effects on respiratory health. Perhaps the events of recent months may also serve to loosen the hold of the Licensed Vintners Association on Irish society, helping the country to move away from the absurdities and disadvantages of its heavy-drinking stereotype. Of added relevance is the evidence of the role of off-license sales during pub closures—inevitable contributors to the social disorder associated with public drinking.

Feasible and enforceable individual-level efforts

Blanket and Immediate Face Mask Policies: Where possible, this might be combined with publicity efforts to de-stigmatize face mask use. The lack of such policies in recent months should be attributed far more to a lack of national preparedness and fear of supply shortages than to any doubts about their effectiveness: even 50% reductions in transmission are well worth the investment, but only if face mask use can be enforced on a routine basis.

Investing in Tech: Contact tracing apps, which are likely to be essential in controlling future community-level outbreaks, should be made accessible to all. But, if adherence to public health guidelines is a government edict, so to should the necessary technology be made available to all: this need not involve the provision of smart phones to every member of the population, but could—in theory—include some kind of low-cost personal and portable alert or monitoring system that does not impinge on privacy. In the same way, efforts to allay fears about personal surveillance implications should, of course, accompany any such step.

Marshaling Volunteers, Businesses, and Communities: Volunteer organization involvement was, for many, inhibited or delayed at critical times in Ireland due to vetting procedures. While such safeguarding efforts are of course essential, their use during times of national public health emergencies should perhaps be reconsidered. In the same way, a mandatory emergency preparedness system for businesses, communities, and individuals—even extending to the predesignation of community wardens—might help to alleviate much of the frustration and powerlessness that many felt.

The above inevitably excludes many other considerations, such as managing economic benefits—should automatic payments avoid the blanket reimbursements and assurances that so troubled the economy via banking guarantees in the past? There is thus a vast amount more that Ireland and many other countries have learned in these regards in recent months; the prospective issue is to ensure that the events of Spring 2020 not be dismissed as one-off anomalies. In an increasingly globalized world, only longer-term efforts to impose health security, ingress and egress polices can ensure that the mistakes of the past are not repeated.

The key distinction that may be easy to overlook in a time of summer weather and declining infection rates is also between epidemic elimination and eradication: while the former may have been achieved, the latter is a significantly different prospect—both in the current case, and for future epidemic preparedness from any infectious disease. Events such as annual lockdown preparedness mobilization and mock tests of national responses should therefore not be dismissed: it is only through rapid preparedness on the individual as well as the national level that future epidemic containment will depend.

Finally, timing may also be critical: as noted above, the immediate application of intense restrictions for shorter time periods, involving entities such as not only the defense forces but also the coast guard and civil defense will inevitably curtail the epidemic in a much more effective way—rather than the gradual phasing up of efforts, which allowed infection to spread in a manner that was, albeit briefly, beyond the control of the government.

Epidemic control equals health security: what developing countries can (still) learn from the global North

Unfortunately, there are only a limited number of ways that data or examples from developed countries are likely to help the developing world, and sub-Saharan Africa in particular, in their preparations for the current global pandemic. There are too many variables in play—not least geographical differences; possibilities of diminution of virulence over time and space; preparation times; available resources; culture; religion; and a host of other possible confounders. Response systems which have shown encouraging quantitative results in one nation may thus be ineffective, or even counterproductive, on other levels and in other places.

Timing, chance and circumstance (such as in the case of Italy) have played key roles in country-by-country impact, meaning that ostensibly positive data cannot always be meaningfully linked to policy, either. Even within Europe, unhelpful comparisons on the interaction between policy and results abound: Sweden, for example, has greater health system capacity than Ireland—and, ultimately, the only point of any social or economic intervention has been to avoid health system overrun. Nonetheless, there are benefits that developing countries may accrue from attention to health security experiences and response strategies in Southeast Asia, Europe, and elsewhere.

Lead time

Much of the developing world is uniquely placed in terms of the current global pandemic. For geographical as much as socio-economic reasons, such countries may have greater time to prepare, and thus—at least theoretically—more opportunities for fine-tuning and tailoring their interventions to local contexts. In many ways, ironically, the developing world is also better equipped for the current situation than Europe or the United States: many populations in sub-Saharan Africa are highly experienced in living with the threats of infectious diseases and associated life expectancy issues, unlike those of more affluent countries.

As such, day-to-day living habits may not be subject to such dramatic change—or, if they are, populations may be more adaptable, based on their past experiences. Although as yet uncertain, the current pandemic may also be linked to affluent pursuits such as international travel and even vaping—neither of which are common in many developing countries. If this is indeed a disease of affluence rather than poverty, then more draconian policies are not going to be cost-effective.

Nonetheless, certain basic policies, without high levels of investment and with a focus on health security, should be advanced. These include health education that both informs and avoids hype and paranoia and associated social unrest and economic disruption; it also relates to the adaptability of epidemic control efforts to different cultural contexts, including religion and political climate.

A further consideration is avoiding information overload—how can developing countries, any more than anywhere else, navigate the increasingly complex global information flows on this subject; when implemented, how can interventions be done in a way that does not disrupt essential services? The flux of fake news in this regard is also a key consideration: developing county governments should not be encouraged to leverage public health efforts for other agendas, as has happened already in many parts of the developed world.

Disease control is thus no excuse for authoritarianism or the limitation of human rights; witness the controversies around internment camps in the 2014 Sierra Leone Ebola outbreak. In the case of South Africa, for example, the conflation of public health and political agendas is open to particular question: the ban on alcohol seems extreme, though logical: the provision of financial assistance only to Black Economic Empowerment companies has less inherent sense to it, and is suggestive of the political advancement of other agendas.

The primacy of health security

A further consideration is the necessity of protecting developing country populations not from the virus, but from international arrivals. Worldwide, the freeze in international travel has been essential, even though (in hindsight) implemented far too late: in the same way, there may be no need to change the day-to-day way of life in developing country townships, as long as residents are protected from interactions with international or regional arrivals.

One further possible consolation is that, if the epidemic were to have been as severe in developing countries as feared, it would likely have happened by now, raising many health security questions regarding ethnicity, culture, and other risk factors that may differ between regions. In this regard, prior epidemics also show us, if nothing else, that adaptable interventions that are suitable to local socio-economic conditions are essential. This includes a focus on utilization of local resources in response design, but also careful consideration of branding in health messaging; making feasible requests that do not deter service utilization, and close attention to a range of other possible sensitivities.

Similarly, ethical considerations related to provision of services for other diseases should be carefully considered, as should the importance of cross-sectoral responses that go beyond ministries of health and utilize resources from a wider range of government departments and public service providers. Should the current pandemic shift to low-level "acceptable risk" endemic status in developing countries, responses also need to be designed with such long-term surveillance and knowledge transfer (as well as emergency response) considerations in mind.

What has worked?

By far the most effective international public health measure, from a health security perspective, has been the closure of borders and limitations of population movements, both internally and between countries. This has worked due to cooperation between transport providers, legislators, and public health experts, but also because of the decisive intervention of politicians in this regard.

Social distancing policies have also been successful, largely because of their humanistic nature—in many cases, the previous status quo of large crowds concentrated in small spaces has been revealed as optional; from over-full commuter trains to other examples of modern overcrowding, few have been revealed to be essential to human progress and survival. The ascendancy of remote working has also been critically important in this regard, though the latter is an option which may not be open to most developing country populations.

Thirdly, the compliance of vulnerable and elderly populations with national directives has been of critical importance. Without this, those at highest risk would likely have rapidly filled many hospital systems to capacity—as was the case in Italy. Optimal measures in developing countries should focus on these lessons learned—with significant adaptation to local conditions. Social distancing efforts are likely to be almost impossible to enforce, for example, while (on the other hand) there is traditionally less mobility by older and vulnerable populations in the developing world, compared to developed countries.

As such, curtailments of travel between countries, districts and regions (in other words, the constriction of relatively loose cordon saintaires or reverse cordon sanitaires) is most likely to be highly effective. In many developing counties, partially enforced borders between counties, districts and regions are already in place: continuing suspension of long-distance travel is both practical, effective, and enforceable in this context.

Similarly, the curtailment of the movement of international populations arriving in to, and living in, developing countries may be of crucial importance. The development of tagging systems for new arrivals to wait out quarantine periods could also be an enforceable initiative in this regard; the further development of effective and efficient "bush telegraph" surveillance information sharing is also likely to be of key importance in curtailing local outbreaks.

A nascent health security checklist for the developing world

Health security is too often considered a purely developed country concern: protecting borders and affluent populations from the arrival of, say, Ebola from West Africa during outbreak periods. This is, perhaps, a misplaced assumption: health security, today, is of greater importance than ever for developing countries, who are now equipped with justifiable rationales for lamenting ingress. In conflict settings such as Sudan, entry by nonnationals in to Khartoum is relatively east compared to the stipulations required for further progress: such policies, once regarded as inherently malign, may have a key emergency response rationale—provided they are not abused. The following may thus help to serve as a draft framework for donor health security assistance, in this regard.

Border Control Support: This might include the provision of fact sheets or other health checks at air and sea ports, as well as donor support of efforts to maintain border security within partner countries.

Utilization of Local Networks: This might occur via many existing foreign assistance initiatives and partners; these networks can play a pivotal role in health education messaging, as well as grassroots surveillance and knowledge transfer. Though donor collaboration with local religious leaders, for example, health education messaging can be effectively integrated in to religious ceremony and teaching.

Health Education: The provision of straightforward, locally adapted health education messaging is likely to be critical. This should not rely on internet promotion, but rather on text messaging, poster campaigns, advertising, and community awareness efforts; radio and popular drama messaging has also met with prior success in this regard. Wherever possible, such initiatives should be integrated with existing health education programs, rather than supplanting them.

Positive Messaging: Combining health protection measures with education on the benefits of limited mobility in the climate change and environmental context might also be of assistance to developing countries—this has been an important incentive for the developed world to comply with emergency regulations.

Xenophobia Avoidance: It may also be important, in developing country health security messaging, to avoid conflation of lockdown efforts with tribalism, regionalism, or other forms of domestic or international xenophobia. Conversely, such policies can be leveraged to promote the opposite: the necessary integration of migrant communities, during periods of limited mobility.

Convertibility: It might also be emphasized in donor–partner dialogs that, strategically, all emergency services should also be convertible to day-to-day use under nonemergency conditions; generalizable into other health condition treatment, but also ready for reactivation. This may help to ensure that white elephant investments that remain unused after the emergency period are avoided.

Home and Community Prevention and Care: In many ways, developed county responses have been based on health system capacity: in that sense, their polices have been as much to protect health systems, as populations. In the developing world, due to a very limited health infrastructure that is often already overwhelmed, the immediate focus should thus be on community-led care and home care strategies.

Prevention and Treatment Campaigns: Information promotion might include the basic measures required, in these regards. In particular, this would include water and sanitation health guidelines; risk factor recommendations such as on smoking; and locally relevant advice such as avoiding the use of indoor fires and other pulmonary and respiratory health initiatives.

Industrial Change Preparations: Industries likely to be highly affected in many developing countries should also be advised on possible consequences at early stages: recommendations on diversifying away from sporting events, tourism, alcohol sales, restaurants, and other services in to more resilient industries may help many to prepare for, or avoid, financial hardship.

Local Ownership: In many cases, government policies, as noted above, may be conflated with other agendas. Thus, in emergency situations, there is a strong case to be made for civil obedience, but also for social authority: trusting local populations to make informed decisions based on local circumstances, rather than exclusive reliance on (often unenforceable) overly paternalistic or controlling efforts that might be construed as didactic or authoritarian.

Reconsidering Protected Health Information: Developing countries may also wish to consider the risks and shortcomings of protected health information in the emergency context. Greater clarity and granularity and resolution in the local control of outbreaks will help with cordon sanitaires, and will likely reduce, rather than increase, damaging stigma and paranoia considerations.

Avoiding Overreliance on Testing: Currently, large-scale testing strategies are unfeasible for the developing world. Associated costs and logistics are likely to continue to be prohibitive for some time—though point-of-entry strategies could be considered. In the HIV/AIDS realm, testing has been of limited practical use in controlling epidemics unless partnered with community referral systems; likewise, contact tracing strategies, particularly when epidemics take place on large scales, are unlikely to be effective in the developing world.

A Focus on Geography: As with many other infectious disease outbreaks, the nature of the current epidemic is inherently urban, with transmission, incidence and prevalence rates all significantly higher in high-density population areas. As such, developing countries may wish to focus efforts, cordon sanitaires, and other border control measures on protecting rural areas (and associated essential food production efforts).

A Focus on Demographics: The current pandemic has been revealed as one which affects elderly and other vulnerable populations with much greater severity that other demographics. Developing countries may wish to make this an inherent part of their response strategies—emphasizing the importance of protecting those population groups, without limiting their basic personal freedoms. Where possible, such efforts should be undertaken from a humanitarian, rather than a health system, perspective.

Urban Aged Populations: Based on a combination of the above two considerations, there may be a particular need for developing countries to focus on protecting vulnerable or elderly populations in urban areas. Though less common in low-resource settings, hospices, care homes, and retirement centers nonetheless remain high risk areas: in this regard, developing countries may still be able to learn from some of the possible oversights of developing countries in this regard

Looking Beyond Treatment or Prevention Paradigms: Too often, epidemic responses are simplistically divided between treatment or prevention. In the current case, treatment options are limited—particularly for severe cases, in the developing world. Even with less severe cases, limited curative or treatment options exist beyond basic recovery measures; likewise, in terms of prevention, no vaccine exists—or is likely to for some time. Similarly, social distancing measures (as noted above)—while effective—are highly unlikely to be effective in many parts of the world due to educational, literacy, and population density considerations.

Prevention and Containment: A key consideration for developing countries is, thus, the difference between prevention and containment. Containment strategies, in geographical terms, are essentially large-scale infection prevention efforts—even at the micro level, they may theoretically reduce risk within restricted areas. With such a community-level focus, geo-containment strategies therefore represent one of the few feasible alternatives to micro-level prevention efforts in developing countries.

Feasible Local Responses: The use of local resources (including recognizing local expertise) is key to epidemic containment in any setting, but particularly in low-resource areas. Donor funding is also likely to be severely curtailed as affluent countries count the cost of their epidemics; thus, as above, the advancement of practical and culturally acceptable measures such as sneezing in to elbow crooks, recommendations on time limits spent in close proximity to others, or the use of home-made face masks may be far more appropriate and effective for the developing world than more socially or resource-demanding recommendations.

Conclusion: a need for bipartisan health security approaches

It is, in all cases, critically important that developing countries avoid over-reacting, or conflating policies with other agendas: to do so is be to undermine public health efforts and erode public trust in authority. Rather, it is vital to steer a line between hype, paranoia, and despair (as characterized by many liberal-democratic responses) versus political or social recklessness (associated with more conservative, libertarian, or authoritarian regimes). Moderate bipartisanism may be the best overarching advice for any developing country—as are considerations of quality of life; extreme health security measures should not, above all, be allowed to excessively or unnecessarily affect other areas of public health, human rights, or economic productivity beyond emergency circumstances.

Selected glossary

Barefoot diplomacy This is used throughout the book as a shorter term for barefoot global health diplomacy: the process by which institutions, individuals, and organizations—no to mention nation states—can leverage the altruistic to achieve the diplomatic in the enlightened self-interest context, thus attracting further resources for infectious disease control.

Epidemics; pandemics; infectious disease control Again, these terms are used largely interchangeably throughout the book. All relate to the fundamental definition of global health used in this book.

Global health This relates, throughout this book, specifically to infectious diseases amongst developing-world populations. However, there are far broader definitions that also apply here: the health of the planet, as well as the containment of infections and epidemics. Thus, global health becomes not just an issue for the world's poor, but also for the developed world.

Internationals This captures the lives and professions of so many of those who travel from their homes with the goal of improving global health in mind. The terms can therefore be applied to consultants, program managers, health attaches, diplomats, businessmen, politicians, doctors, nurses, or other paramedical services or advisors.

Locals This attempts to capture all those who, in epidemic or public health emergency or other situations, are the ones who live and work in the places most affected. By implication, locals are often from the poorest countries in the world.

Program, project, intervention, response These terms are used interchangeably throughout the book. My apologies to those who prefer that they be used distinctly: in my experience, there is often very little to differentiate the terms in practice. Together, they encompass all field efforts to conduct health programs, around the world.

Public health emergencies Often used in an *ad hoc* way throughout this book; it deals with the many occasions in which epidemics and infectious diseases are either the cause or the effect of natural disasters or other political or economic effects, and thereby falls under the broader terms of global health as well.

Smart The term smart global health has been coined by the Center for Strategic and International Studies (CSIS). I acknowledge their thinking and use of this term, and the inspiration it has provided for many parts of this book. Essentially, smart is here defined in the same way as a smart phone: a multifariousness that allows global health efforts to do more than would have traditionally been expected of them.

Index

Note: Page numbers followed "*f*" and "*b*" refer to figures and boxes, respectively.

Printed in the United States
By Bookmasters